Principles and Standards for School Mathematics Navigations Series

Navigating through Probability in Grades 6–8

George W. Bright
Dargan Frierson, Jr.
James E. Tarr
Cynthia Thomas

Susan Friel
Grades 6–8 Editor
Peggy A. House
Navigations Series Editor

NATIONAL COUNCIL OF
TEACHERS OF MATHEMATICS

Copyright © 2003 by
The National Council of Teachers of Mathematics, Inc.
1906 Association Drive, Reston, VA 20191-1502
(703) 620-9840; (800) 235-7566
www.nctm.org

All rights reserved

ISBN 0-87353-523-5

Permission to photocopy limited material from *Navigating through Probability in Grades 6–8* is granted for educational purposes. Permission must be obtained when content from this publication is used commercially, when the material is quoted in advertising, when portions are used in other publications, or when charges for copies are made. The use of material from *Navigating through Probability in Grades 6–8*, other than in those cases described, should be brought to the attention of the National Council of Teachers of Mathematics.

The contents of the CD-ROM may not be reproduced, distributed, or adapted without the written consent of NCTM, except as noted here: The blackline masters may be downloaded and reproduced for classroom distribution; the applets may be used for instructional purposes in one classroom at a time.

The publications of the National Council of Teachers of Mathematics present a variety of viewpoints. The views expressed or implied in this publication, unless otherwise noted, should not be interpreted as official positions of the Council.

Printed in the United States of America

Table of Contents

About This Book vii
Introduction 1

Chapter 1
The Concept of Probability 11
    Who Will Win? 17
    Two Dice 20
    More Often and Most Often 23
    Empirical Probabilities 26

Chapter 2
Probability Distributions 29
    Fair Spinners 38
    Dice Differences 41
    Test Guessing 43
    Number Golf 45
    First Head 47
    Strings of Heads 49

Chapter 3
Prediction and the Law of Large Numbers 51
    Racing Game 56
    Two Hospitals 61
    Gumball Machine 64
    How Will It Land? 67
    The Long Flight Home 69

Chapter 4
Connecting Probability and Statistics 73
    Dixie's Basketball Contest 78
    Newspaper Route 80
    How Black Is a Zebra? 84

Looking Back and Looking Ahead 87

Appendix
Blackline Masters and Solutions 89
    Who Will Win? 90
    Two Dice 91
    Pets 93
    More Often and Most Often 94
    Ratios 95
    Fair Spinners 99
    Dice Differences 102
    Test Guessing 104
    Number Golf: One Die 105

Number Golf: Two Dice 106
First Head 107
Strings of Heads 109
Racing Game 113
Pooled Racing-Game Data 115
Two Hospitals 116
Gumball Machine 119
How Will It Land? 121
The Long Flight Home 123
The Long Flight Home—Extension 124
Dixie's Basketball Contest 125
Newspaper Route 127
How Black Is a Zebra? 128
Solutions for the Blackline Masters 130

REFERENCES 139

CONTENTS OF CD-ROM

Introduction

Table of Standards and Expectations, Data Analysis and Probability, Pre-K–12

Applet Activities
Spinner
Adjustable Spinner
Racing Game with Two Dice

Blackline Masters and Template
Blackline Masters
Half-Centimeter Grid Paper

Readings from Publications of the National Council of Teachers of Mathematics

Enriching Students' Mathematical Intuitions with Probability Games and Tree Diagrams
    Leslie Aspinwall and Kenneth L. Shaw
    Mathematics Teaching in the Middle School

Counting Attribute Blocks: Constructing Meaning for the Multiplication Principle
    Elliott Bird
    Mathematics Teaching in the Middle School

Genetics as a Context for the Study of Probability
    Daniel J. Brahier
    Mathematics Teaching in the Middle School

Fair Games, Unfair Games
    George W. Bright, John G. Harvey, and Margariete Montague Wheeler
    Teaching Statistics and Probability

Readers Write: Is Rock, Scissors, and Paper a Fair Game?
    Donald J. Dessart and Charlene M. DeRidder
    Mathematics Teaching in the Middle School

Probability on a Budget
    William A. Ewbank and John L. Ginther
    Mathematics Teaching in the Middle School

Roll the Dice: An Introduction to Probability
    Andrew Freda
    Mathematics Teaching in the Middle School

Understanding Students' Probabilistic Reasoning
    Graham A. Jones, Carol A. Thornton, Cynthia W. Langrall, and
    James E. Tarr
    Developing Mathematical Reasoning in Grades K–12

Push-Penny: What Is Your Expected Score?
    Gary Kader and Mike Perry
    Mathematics Teaching in the Middle School

From *The Giver* to *The Twenty-One Balloons:* Explorations with Probability
    Ann Lawrence
    Mathematics Teaching in the Middle School

Socrates and the Three Little Pigs: Connecting Patterns, Counting Trees, and Probability
    Denisse R. Thompson and Richard A. Austin
    Mathematics Teaching in the Middle School

Racing to Understand Probability
    Laura R. Van Zoest and Rebecca K. Walker
    Mathematics Teaching in the Middle School

Exploring Probability through an Evens-Odds Dice Game
    Lynda R. Wiest and Robert J. Quinn
    Mathematics Teaching in the Middle School

About This Book

Historically, probability has not been treated coherently in most curriculum materials for school mathematics, in part because although the ideas of probability are intuitively accessible, they are mathematically complex. Consider some common examples: meteorologists may predict "a 30 percent chance of showers"; each year, the media report several high-profile stories about winners of lotteries with large payoffs; and researchers warn the public about the increased risk of disease from eating particular foods or being exposed to various environmental conditions. We can live very fulfilling lives with only an intuitive sense of the probabilities in these stories, but interpreting the stories fully requires an understanding of probability. To understand probability deeply, we need access to a wide variety of mathematical tools, representations, and forms of argument that are not necessarily typical of other areas of mathematics. Working with probability concepts throughout the school years is a necessary part of building a deep understanding of probability.

Principles and Standards for School Mathematics (National Council of Teachers of Mathematics [NCTM] 2000) directly addresses probability in one of the four Data Analysis and Probability Standards for the middle grades. The expectations for students of probability in grades 6–8 include using appropriate terminology, testing conjectures about the results of experiments and simulations, and using a variety of methods to compute probabilities (NCTM 2000, p. 248). This list reveals the complexity of this topic and suggests some ways that probability can be connected to other mathematical ideas.

Probabilistic thinking, perhaps more than any other kind of mathematical thinking, is influenced by events in our everyday lives. For example, it is well known (see, e.g., Kahneman, Slovic, and Tversky [1982]) that recent events (e.g., knowing someone who has recently been called by a pollster) often cause people to inflate their notions of the likelihood of a similar event (e.g., being called by a pollster). Because of the prevalence of these real-world interferences, it is difficult for teachers to counter faulty intuitive reasoning that leads students astray (Konold 1989). Also, in ordinary language, we use words related to probability (e.g., *chance*, *unlikely*) in ways that may interfere with students' development of the more precise mathematical meanings of the terms. When the forecast chance of rain is any value less than 50 percent, we might say, "It is unlikely that it will rain." But classifying such a wide range of values as *unlikely* may interfere with the development of the notion *equally likely* in mathematics. Teachers need to be alert to students' discussions of probability problems in order to detect the interference of faulty intuitive understandings with the development of more-precise, correct mathematical understandings.

Understand and apply basic concepts of probability

Research on how probabilistic thinking develops is not complete, but several publications (e.g., Fischbein [1975]; Piaget and Inhelder [1975]; Shaughnessy [1992]) reveal elements of how it develops. These studies offer a framework for teachers to use in determining what their students know. Asking students to explain their thinking, however, seems the most direct way to discover how students are understanding and misunderstanding the important ideas of probability.

The Organization of This Book

This book consists of four chapters: "The Concept of Probability," "Probability Distributions," "Prediction and the Law of Large Numbers," and "Connecting Probability and Statistics." Some of the essential ideas (beyond basic definitions) about probability are presented in the activities and discussions in this book. One important idea is that when experiments can be repeated, probabilities can be approximated by computing the relative frequencies of pooled data. In order for such approximations to be believable, it is important that the experiments be repeated under identical conditions. If experiments are conducted under different conditions, the results cannot be combined.

Some events can be studied only empirically; it may not be possible to determine theoretical probabilities for such events as a thumbtack landing point up or point down (i.e., resting on its side). The best we can hope to determine is an approximate, experimental probability of "point down," although the probability will depend on the particular thumbtack being tossed.

Related to the idea that probabilities can be determined experimentally is the notion that outcomes from experiments with large, representative samples must be taken seriously when they differ from predictions. Unexpected experimental evidence should act as a challenge to students' intuitive thinking about a situation. Getting students (or adults, for that matter) even to consider the possibility that their intuitive thinking might be inaccurate is not easy. But it is important for the development of sound mathematical thinking.

A second important idea is the representativeness of samples. Individual events and small samples are often not representative of trends. For example, we cannot expect that ten tosses of a coin will reveal enough information about the coin to decide if it is biased. Data from several thousand tosses, though, can be expected to be more representative of a trend. This fact supports the idea that it is appropriate to pool data by, for example, having every student toss the same coin ten times and then tallying all the class's data. When data are pooled, individual students' samples (and, thus, their distributions) are no longer identifiable. Students need to understand that their contributions to the pooled data are important, even though they cannot identify exactly what part of the data they generated. Teachers can help students appreciate their role by explaining that the class distribution has greater credibility than an individual's distribution. For example, some student is fairly likely to have a series of coin tosses that yield eight heads in ten tosses. Although the student's contribution is important, we would not expect that in the class's pooled sample, eighty heads would result from one hundred tosses.

A third important idea is that events with small probabilities *can* occur. A classic example is state lotteries. The likelihood of any particular person's winning a lottery is extremely small. Yet many people are willing to play against the odds. Students should come to understand that winning a big payoff, although unlikely, is possible and that some people are willing to take the considerable risk of losing a bet in the hope of attaining an extremely favorable outcome.

Each chapter in this book begins with a discussion of the important probability ideas in the chapter. Then a "preassessment" activity is presented that helps teachers understand what knowledge their students might already have about these ideas. The preassessment is followed by other activities. Many of the activities have blackline masters that can be used directly with students. The blackline masters, signaled by an icon, can be found in the appendix, along with solutions to the problems where appropriate. The masters can also be printed from the CD-ROM that accompanies the book. The CD-ROM, also signaled by an icon, includes resources for professional development and three applets that provide an alternative way of engaging students with some of the same probability ideas presented in the book. Throughout the book, an icon appears in the margin next to references to *Principles and Standards for School Mathematics*.

Instruction in Probability

Teachers must plan instruction that helps students connect the ideas of probability and then connect those ideas with other mathematical ideas (e.g., statistics). Students should be given many opportunities in many different settings to think probabilistically; for example, gathering data by rolling number cubes, tossing coins, and drawing chips out of a bag are more likely collectively to support important learning than experiences with only one of those methods. Students take a long time to develop clear notions of chance, and in order to generalize correctly about complex ideas, they need to be exposed to many examples (and nonexamples) of each of the main ideas of probability. A variety of methods and tools, such as organized lists, tree diagrams, and area models, should also be used for solving problems.

Students are more likely to connect their understanding of probability with their everyday lives if they are asked to reason probabilistically about data rather than simply to provide answers to questions. Students need to develop probabilistic thinking by using the data they generate in experiments as well as data from other sources. Historically, greater attention has been given in school curricula to analyzing theoretical distributions—for example, the six-by-six chart of possible dice rolls—than to analyzing data. Current educational trends, however, favor activities that ask students to reason from data—for example, data gathered by rolling dice (e.g., Racing Game in chapter 3). Both approaches are illustrated in the activities in this book.

Simulations should also be a part of probability instruction in the middle grades. For example, if we assume that boys and girls are equally likely to be born, we can simulate births by flipping a fair coin. Students need to understand, however, that for a simulation to be helpful, the characteristics of the simulation must accurately model the conditions of the situation. (For example, male and female babies are not exactly equally likely, but the likelihoods may be close enough for a coin-flipping simulation to be useful.)

If technology is used to generate a simulation, students can have access to very large samples very quickly or to multiple large samples that reflect different assumptions about the situation. But what do students understand about the data generated by a simulation? Do they

Key to Icons

Principles and Standards

Blackline Master

CD-ROM

Three different icons appear in the book, as shown in the key. One alerts readers to material quoted from *Principles and Standards for School Mathematics,* another points them to supplementary materials on the CD-ROM that accompanies the book, and a third signals the blackline masters and indicates their locations in the appendix.

consider those data to be as accurate as the data gathered by actually flipping a coin? Very little information is available about students' beliefs about technology-generated data. Teachers need to probe students' thinking about this issue to reveal any biases and misconceptions that students may be harboring. If they are suspicious about computer-generated data, the teacher may want to revert to using physical manipulatives.

One caution is that electronic technology (e.g., graphing calculators, computers) should not completely replace such "less sophisticated" methods as a random-number table or rolls of the dice. Less sophisticated devices are very useful—and often very effective—for helping students understand the underlying mathematical processes, even though such devices may limit the range of situations that can be investigated. Sophisticated technology may mask processes and fail to encourage clear thinking.

NAVIGATING *through* PROBABILITY

GRADES 6–8

Introduction

The Data Analysis and Probability Standard in *Principles and Standards for School Mathematics* (NCTM 2000) is an affirmation of a fundamental goal of the mathematics curriculum: to develop critical thinking and sound judgment based on data. These skills are essential not only for a select few but for every informed citizen and consumer. Staggering amounts of information confront us in almost every aspect of contemporary life, and being able to ask good questions, use data wisely, evaluate claims that are based on data, and formulate defensible conclusions in the face of uncertainty have become basic skills in our information age.

In working with data, students encounter and apply ideas that connect directly with those in the other strands of the mathematics curriculum as well as with the mathematical ideas that they regularly meet in other school subjects and in daily life. They can see the relationship between the ideas involved in gathering and interpreting data and those addressed in the other Content Standards—Number and Operations, Algebra, Measurement, and Geometry—as well as in the Process Standards—Reasoning and Proof, Representation, Communication, Connections, and Problem Solving. In the Navigations series, the *Navigating through Data Analysis and Probability* books elaborate the vision of the Data Analysis and Probability Standard outlined in *Principles and Standards*. These books show teachers how to introduce important statistical and probabilistic concepts, how the concepts grow, what to expect students to be able to do and understand during and at the end of each grade band, and how to assess what they know. The books also introduce representative instructional activities that help translate the vision of *Principles and Standards* into classroom practice and student learning.

Fundamental Components of Statistical and Probabilistic Thinking

Principles and Standards sets the Data Analysis and Probability Standard in a developmental context. It envisions teachers as engaging students from a very young age in working directly with data, and it sees this work as continuing, deepening and growing in sophistication and complexity as the students move through school. The expectation is that all students, in an age-appropriate manner, will learn to—

- formulate questions that can be addressed with data and collect, organize, and display relevant data to answer them;
- select and use appropriate statistical methods to analyze data;
- develop and evaluate inferences and predictions that are based on data; and
- understand and apply basic concepts of probability.

Formulating questions that can be addressed with data and collecting, organizing, and displaying relevant data to answer them

No one who has spent any time at all with young children will doubt that they are full of questions. Teachers of young children have many opportunities to nurture their students' innate curiosity while demonstrating to them that they themselves can gather information to answer some of their questions.

At first, children are primarily interested in themselves and their immediate surroundings, and their questions center on such matters as "How many children in our class ride the school bus?" or "What are our favorite flavors of ice cream?" Initially, they may use physical objects to display the answers to their questions, such as a shoe taken from each student and placed appropriately on a graph labeled "The Kinds of Shoes Worn in Kindergarten." Later, they learn other methods of representation using pictures, index cards, sticky notes, or tallies. As children move through the primary grades, their interests expand outward to their surroundings, and their questions become more complex and sophisticated. As that happens, the amount of collectible data grows, and the task of keeping track of the data becomes more challenging. Students then begin to learn the importance of framing good questions and planning carefully how to gather and display their data, and they discover that organizing and ordering data will help uncover many of the answers that they seek. However, learning to refine their questions, planning effective ways to collect data, and deciding on the best ways to organize and display data are skills that children develop only through repeated experiences, frequent discussions, and skillful guidance from their teachers. By good fortune, the primary grades afford many opportunities—often in conjunction with lessons on counting, measurement, numbers, patterns, or other school subjects—for children to pose interesting questions and develop ways of collecting data that will help them formulate answers.

As students move into the upper elementary grades, they will continue to ask questions about themselves and their environment, but their questions will begin to extend to their school or the community or the world beyond. Sometimes, they will collect their own data; at other times, they will use existing data sets from a variety of sources. In either case, they should learn to exercise care in framing their questions and determining what data to collect and when and how to collect them. They should also learn to recognize differences among data-gathering techniques, including observation, measurement, experimentation, and surveying, and they should investigate how the form of the questions that they seek to answer helps determine what data-gathering approaches are appropriate. During these grades, students learn additional ways of representing data. Tables, line plots, bar graphs, and line graphs come into play, and students develop skill in reading, interpreting, and making various representations of data. By examining, comparing, and discussing many examples of data sets and their representations, students will gain important understanding of such matters as the difference between categorical and numerical data, the need to select appropriate scales for the axes of graphs, and the advantages of different data displays for highlighting different aspects of the same data.

During middle school, students move beyond asking and answering the questions about a single population that are common in the earlier years. Instead, they begin posing questions about relationships among several populations or samples or between two variables within a single population. In grades 6–8, students can ask questions that are more complex, such as "Which brand of laundry detergent is the best buy?" or "What effect does light [or water or a particular nutrient] have on the growth of a tomato plant?" They can design experiments that will allow them to collect data to answer their questions, learning in the process the importance of identifying relevant data, controlling variables, and choosing a sample when it is impossible to collect data on every case. In these middle school years, students learn additional ways of representing data, such as with histograms, box plots, or relative-frequency bar graphs, and they investigate how such displays can help them compare sets of data from two or more populations or samples.

By the time students reach high school, they should have had sufficient experience with gathering data to enable them to focus more precisely on such questions of design as whether survey questions are unambiguous, what strategies are optimal for drawing samples, and how randomization can reduce bias in studies. In grades 9–12, students should be expected to design and evaluate surveys, observational studies, and experiments of their own as well as to critique studies reported by others, determining if they are well designed and if the inferences drawn from them are defensible.

Selecting and using appropriate statistical methods to analyze data

Teachers of even very young children should help their students reflect on the displays that they make of the data that they have gathered. Students should always thoughtfully examine their representations to determine what information they convey. Teachers can prompt young children to derive information from data displays through questions like "Do more children in our class prefer vanilla ice cream, or do more prefer chocolate ice cream?" As children try to interpret their work, they come to realize that data must be ordered and organized to convey answers to their questions. They see how information derived from data, such as their ice cream preferences, can be useful—in deciding, for example, how much of particular flavors to buy for a class party. In the primary grades, children ordinarily gather data about whole groups—frequently their own class—but they are mainly interested in individual data entries, such as the marks that represent their own ice cream choices. Nevertheless, as children move through the years from prekindergarten to grade 2, they can be expected to begin questioning the appropriateness of statements that are based on data. For example, they may express doubts about such a statement as "Most second graders take ballet lessons" if they learn that only girls were asked if they go to dancing school. They should also begin to recognize that conclusions drawn about one population may not apply to another. They may discover, for instance, that bubble gum and licorice are popular ice cream flavors among their fellow first graders but suspect that this might not necessarily be the case among their parents.

In contrast with younger children, who focus on individual, often personal, aspects of data sets, students in grades 3–5 can and should be guided to see data sets as wholes, to describe whole sets, and to compare one set with another. Students learn to do this by examining different sets' characteristics—checking, for example, values for which data are concentrated or clustered, values for which there are no data, or values for which data are unusually large or small (*outliers*). Students in these grades should also describe the "shape" of a whole data set, observing how the data spread out to give the set its *range*, and finding that range's center. In grades 3–5, the center of interest is in fact very often a measure of a data set's center—the *median* or, in some cases, the *mode*. In the process of learning to focus on sets of data rather than on individual entries, students should start to develop an understanding of how to select *typical* or *average* (*mean*) values to represent the sets. In examining similarities and differences between two sets, they should explore what the means and the ranges tell about the data. By using standard terms in their discussions, students in grades 3–5 should be building a precise vocabulary for describing the characteristics of the data that they are studying.

By grade 5, students may begin to explore the concept of the mean as a balance point in an informal way, but a formal understanding of the mean and its use in describing data sets does not become important until grades 6–8. By this time, just being able to compute the mean is no longer enough. Students need ample opportunities to develop a fundamental conceptual understanding—for example, by comparing the usefulness and appropriateness of the mean, the median, and the mode as ways of describing data sets in different contexts. In middle school, students should also explore questions that are more probing, such as "What impact does the spread of a distribution have on the value of the mean [or the median]?" Or "What effect does changing one data value [or more than one] have on different measures of center—the mean, the median, and the mode?" Technology, including spreadsheet software, calculators, and graphing software, becomes an important tool in grades 6–8, enabling students to manipulate and control data while they investigate how changes in certain values affect the mean, the median, or the distribution of a set of data. Students in grades 6–8 should also study important characteristics of data sets, such as *symmetry, skewness,* and *interquartile range,* and should investigate different types of data displays to discover how a particular representation makes such characteristics more or less apparent.

As these students move on into grades 9–12, they should grow in their ability to construct an appropriate representation for a set of univariate data, describe its shape, and calculate summary statistics. In addition, high school students should study linear transformations of univariate data, investigating, for example, what happens if a constant is added to each data value or if each value is multiplied by a common factor. They should also learn to display and interpret bivariate data and recognize what representations are appropriate under particular conditions. In situations where one variable is categorical—for example, gender—and the other is numerical—a measurement of height, for instance—students might use appropriately paired box plots or histograms to compare the heights of males and females in a given group. By contrast, students who are presented with bivariate numerical data—for example, measurements of height and arm span—might use a scatterplot to represent their data, and they should be able to describe the shape of the scatterplot and use it to analyze the relationship between the two lengths measured—height and arm span. Types of analyses expected of high school students include finding functions that approximate or "fit" a scatterplot, discussing different ways to define "best fit," and comparing several functions to determine which is the best fit for a particular data set. Students should also develop an understanding of new concepts, including *regression, regression line, correlation,* and *correlation coefficient.* They should be able to explain what each means and should understand clearly that a correlation is not the same as a causal relationship. In grades 9–12, technology that allows users to plot, move, and compare possible regression lines can help students develop a conceptual understanding of residuals and regression lines and can enable them to compute the equation of their selected line of best fit.

Developing and evaluating inferences and predictions that are based on data

Observing, measuring, or surveying every individual in a population is an appropriate way of gathering data to answer selected questions. Such "census data" is all that we expect from very young children, and teachers in the primary grades should be content when their students confine their data gathering and interpretation to their own class or another small group. But as children mature, they begin to understand that a principal reason for gathering and analyzing data is to make inferences and predictions that apply beyond immediately available data sets. To do that requires sampling and other more advanced statistical techniques.

Teachers of young children lay a foundation for later work with inference and prediction when they ask their students whether they think that another group of students would get the same answers from data that they did. After discussing the results of a survey to determine their favorite books, for example, children in one first-grade class might conclude that their peers in the school's other first-grade class would get similar results but that the fourth graders' results might be quite different. The first graders could speculate about why this might be so.

As students move into grades 3–5, they should be expected to expand their ability to draw conclusions, make predictions, and develop arguments based on data. As they gain experience, they should begin to understand how the data that they collect in their own class or school might or might not be representative of a larger population of students. They can begin to compare data from different samples, such as several fifth-grade classes in their own school or other schools in their town or state. They can also begin to explore whether or not samples are representative of the population and identify factors that might affect representativeness. For example, they could consider a question like "Would a survey of children's favorite winter sports get similar results for samples drawn from Colorado, Hawaii, Texas, and Ontario?" Students in the upper grades should also discuss differences in what data from different samples show and factors that might account for the observed results, and they can start developing hypotheses and designing investigations to test their predictions.

It is in the middle grades, however, that students learn to address matters of greater complexity, such as the relationship between two variables in a given population or sample, or the relationships among several populations or samples. Two concepts that are emphasized in grades 6–8 are *linearity* and *proportionality*, both of which are important in developing students' ability to interpret and draw inferences from data. By using scatterplots to represent paired data from a sample—for example, the height and stride length of middle schoolers—students might observe whether the points of the scatterplot approximate a line, and if so, they can attempt to draw the line to fit the data. Using the slope of that line, students can make conjectures about a relationship between height and stride length. Furthermore, they might decide to compare a scatterplot for middle school boys with one for middle school girls to determine if a similar ratio applies for both groups. Or they might draw box plots or relative-frequency histograms to represent data on the heights of samples of middle school boys and high school boys to investigate the variability in height of boys of different ages. With the help of graphing technology, students can examine many data sets and learn to differentiate between linear and nonlinear relationships, as well as to recognize data sets that exhibit no relationship at all. Whenever possible, they should attempt to describe observed relationships mathematically and discuss whether the conjectures that they draw from the sample data might apply to a larger population. From such discussions, students can plan additional investigations to test their conjectures.

As students progress to and through grades 9–12, they can use their growing ability to represent data with regression lines and other mathematical models to make and test predictions. In doing so, they learn that inferences about a population depend on the nature of the samples, and concepts such as *randomness, sampling distribution*, and *margin of error* become important. Students will need firsthand experience with many different statistical examples to develop a deep understanding of the powerful ideas of inference and prediction. Often that experience will come through simulations that enable students to perform hands-on experiments while developing a more intuitive understanding of the relationship between characteristics of a sample and the corresponding characteristics of the population from which the sample was drawn. Equipped with the concepts learned through simulations, students can then apply their understanding by analyzing statistical inferences and critiquing reports of data gathered in various contexts, such as product testing, workplace monitoring, or political forecasting.

Understanding and applying basic concepts of probability

Probability is connected to all mathematics from number to geometry. It has an especially close connection to data collection and analysis. Although students are not developmentally ready to study probability in a formal way until much later in the curriculum, they should begin to lay the foundation for that study in the years from prekindergarten to grade 2. For children in these early years, this means informally considering ideas of likelihood and chance, often by thinking about such questions as "Will it be warm tomorrow?" and realizing that the answer may depend on particular conditions, such as where they live or what month it is. Young children also recognize that some things are sure to happen whereas others are impossible, and they begin to develop notions of *more likely* and *less likely* in various everyday contexts. In addition, most children have experience with common devices of chance used in games, such as spinners and dice. Through hands-on experience, they become aware that certain numbers are harder than others to get with two dice and that the pointer on some spinners lands on certain colors more often than on others.

In grades 3–5, students can begin to think about probability as a measurement of the likelihood of an event, and they can translate their earlier ideas of *certain, likely, unlikely,* or *impossible* into quantitative representations using 1, 0, and common fractions. They should also think about events that are neither certain nor impossible, such as getting a 6 on the next roll of a die. They should begin to understand that although they cannot know for certain what will happen in such a case, they can associate with the outcome a fraction that represents the frequency with which they could expect it to occur in many similar situations. They can also use data that they collect to estimate probability—for example, they can use the results of a survey of students' footwear to predict whether the next student to get off the school bus will be wearing brown shoes.

Students in grades 6–8 should have frequent opportunities to relate their growing understanding of proportionality to simple probabilistic situations from which they can develop notions of chance. As they refine their understanding of the chance, or likelihood, that a certain event will occur, they develop a corresponding sense of the likelihood that it will not occur, and from this awareness emerge notions of complementary events, mutually exclusive events, and the relationship between the probability of an event and the probability of its complement. Students should also investigate simple compound events and use tree diagrams, organized lists, or similar descriptive methods to determine probabilities in such situations. Developing students' understanding of important concepts of probability—not merely their ability to compute probabilities—should be the teacher's aim. Ample experience is important, both with hands-on experiments that generate empirical data and with computer simulations that produce large data samples. Students should then apply their understanding of probability and proportionality to make and test conjectures about various chance events, and they should use simulations to help them explore probabilistic situations.

Concepts of probability become increasingly sophisticated during grades 9–12 as students develop an understanding of such important ideas as *sample space*, *probability distribution*, *conditional probability*, *dependent* and *independent events*, and *expected value*. High school students should use simulations to construct probability distributions for sample spaces and apply their results to predict the likelihood of events. They should also learn to compute expected values and apply their knowledge to determine the fairness of a game. Teachers can reasonably expect students at this level to describe and use a sample space to answer questions about conditional probability. The solid understanding of basic ideas of probability that students should be developing in high school requires that teachers show them how probability relates to other topics in mathematics, such as counting techniques, the binomial theorem, and the relationships between functions and the area under their graphs.

Developing a Data Analysis and Probability Curriculum

Principles and Standards reminds us that a curriculum that fosters the development of statistical and probabilistic thinking must be coherent, focused, and well articulated—not merely a collection of lessons or activities devoted to diverse topics in data analysis and probability. Teachers should introduce rudimentary ideas of data and chance deliberately and purposefully in the early years, deepening and expanding their students' understanding of them through frequent experiences and applications as students progress through the curriculum. Students must be continually challenged to learn and apply increasingly sophisticated statistical and probabilistic thinking and to solve problems in a variety of school, home, and real-life settings.

The six *Navigating through Data Analysis and Probability* books make no attempt to present a complete, detailed data analysis and probability curriculum. However, taken together, these books illustrate how selected "big ideas" behind the Data Analysis and Probability Standard develop this strand of the mathematics curriculum from prekindergarten through grade 12. Many of the concepts about data analysis and probability that the books present are closely tied to topics in algebra, geometry, number, and measurement. As a result, the accompanying activities, which have been especially designed to put the Data Analysis and Probability Standard into practice in the classroom, can also reinforce and enhance students' understanding of mathematics in the other strands of the curriculum, and vice versa.

Because the methods and ideas of data analysis and probability are indispensable components of mathematical literacy in contemporary life, this strand of the curriculum is central to the vision of mathematics education set forth in *Principles and Standards for School Mathematics*. Accordingly, the *Navigating through Data Analysis and Probability* books are offered to educators as guides for setting successful courses for the implementation of this important Standard.

NAVIGATIONS SERIES

GRADES 6–8

NAVIGATING through PROBABILITY

Chapter 1
The Concept of Probability

Important Mathematical Ideas

When students enter the middle grades, they will almost certainly have had some experiences with the ideas of probability. They might, for example, be able to talk about some events that are certain to happen (e.g., the sun will rise in the east), some events that are certain not to happen (e.g., a green elephant will walk through the classroom in the next five minutes), and some events that may or may not happen (e.g., it will rain tomorrow). It is important to build on that basic knowledge to develop a more sophisticated understanding of the concept of probability.

Fundamental Ideas about Probability

A *probability* is a number between 0 and 1 that measures the likelihood of an event. A probability of 0 indicates that the event will not happen, and a probability of 1 indicates that the event is certain to happen. The probability of an event can be estimated by the fraction of time that the event will happen over the long term; that is, if there are many situations in which the event might happen, the probability indicates the expected fraction of these situations in which the event will happen. For example, the probability of rolling one die and getting a 3 is 1/6; we expect 3 to occur about one-sixth of the time in a large number of rolls. It is important for students to understand that a probability does not apply to each situation but rather to the expectations associated with many potential repetitions of the event.

> An *event* is the outcome of a trial. A *simple event* (usually called an *event*) is a single outcome. A *compound event* is an event that consists of more than one outcome. For example, when a die is rolled, the outcome "six on the top face" is a simple event, whereas the outcome "prime number on the top face" is a compound event, since any of the numbers 2, 3, or 5 is a prime number.

Students gain the greatest benefit from instruction in probability when they have a firm understanding of fraction concepts, especially the ordering of fractions. This is not to say that teachers should delay instruction in probability until students have mastered fractions. Rather, they should be alert during probability instruction for indications that students are, or might be, misunderstanding some important element of ordering fractions (e.g., knowing that 3/4 < 7/8). If teachers detect misconceptions, they might choose to address those misconceptions separately.

Fundamental to understanding probability is skill at specifying the sample space for an event. A *sample space* is a specification (e.g., list, table) of all the events that might happen in a given random experiment. For example, if we toss a coin and record what happens after the coin lands, we might record a head, a tail, or an edge. The chance of the coin's landing on its edge, however, is so remote that we often ignore that option and consider only a head and a tail. The sample space for one toss of a coin, then, can be represented as {H, T}. Similarly, the sample space for the roll of one die or number cube can be represented as {1, 2, 3, 4, 5, 6}. These sample spaces are probably quite familiar to middle-grades students.

Ewbank and Ginther (2002) discuss some ways of helping students understand sample spaces and probabilities for dice numbered in various ways.

Determining sample spaces for repeated trials is often challenging for students. A simple situation that can help students understand some of the complexities is counting heads for the tossing of two coins. Most students will readily admit that if they toss two coins, they might get no heads, one head, or two heads, but they are much less sure about whether these three outcomes are or are not equally likely to occur—that is, whether these three outcomes do or do not have the same probability. If a teacher asks students to list the outcomes of tossing two coins, some will write something like the following:

    0 heads,   2 tails
    1 head,   1 tail
    2 heads,   0 tails

This list is misleading, since two of the events—(0 heads, 2 tails) and (2 heads, 0 tails)—are simple, whereas one event—(1 head, 1 tail)—is compound. One way to understand the compound event is to let the two coins be different—for example, one penny and one nickel. The event (1 head, 1 tail) is composed of two simple events: (head on penny, tail on nickel) and (tail on penny, head on nickel). "Exactly one head" can happen in two ways, so its probability should be twice as great as the probability of either of the two simple events. Thinking of the two coins as penny and nickel also helps students create a complete list of the four equally likely outcomes:

- Tail on penny, tail on nickel
- Tail on penny, head on nickel
- Head on penny, tail on nickel
- Head on penny, head on nickel

"Middle-grades students should learn and use appropriate terminology and should be able to compute probabilities for simple compound events, such as the number of expected occurrences of two heads when two coins are tossed 100 times." (NCTM 2000, p. 51)

The sample space could also be represented by a tree diagram (see fig. 1.1). Of the four possible simple events, two of them together form the event "exactly one head." The probability of exactly one head on the toss of two coins is 2/4, or 1/2, whereas the probability of zero

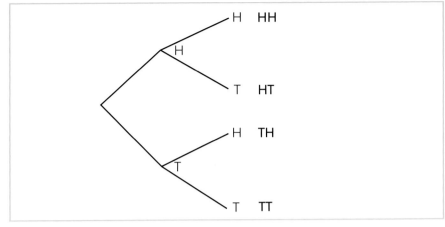

Fig. **1.1.**

A tree diagram of the possible outcomes of tossing two coins

heads on the toss of two coins is 1/4 and the probability of two heads on the toss of two coins is 1/4. The four simple outcomes are equally likely, but the likelihood of zero heads or two heads is not equal to that of one head.

Teachers can extend this idea to the tossing of three coins (see fig. 1.2). Eight simple, equally likely outcomes are possible; in this example, three of them together form the event "exactly one head." The probability of exactly one head on the toss of three coins is 3/8. It is important that students investigate why there are eight possibilities for the tossing of three coins. Since two outcomes are possible for each coin and since three coins are being tossed, there are $2 \times 2 \times 2 = 8$ possible values. This example demonstrates the mathematically important *counting principle*.

Another way to represent probabilities is with geometric regions, such as a square. For example, if a square represents all outcomes, then a square divided into two congruent regions can represent head and tail for one toss of a coin (see fig. 1.3). If each of these subregions is divided into two congruent regions, the four regions can represent the four outcomes for tossing two coins: HH, HT, TH, and TT. Dividing each of these four regions in half yields eight regions that show the outcomes

Aspinwall and Shaw (2000) and Thompson and Austin (1999) give more information about how to use tree diagrams in understanding the counting principle and exploring probability, and Bird (2000) discusses ways of teaching the counting principle.

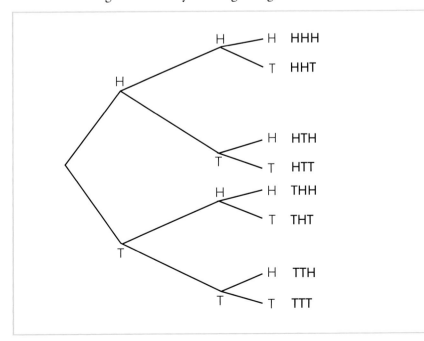

Fig. **1.2.**

A tree diagram of the possible outcomes of tossing three coins of tossing three coins: HHH, HHT, HTH, HTT, THH, THT, TTH, and TTT. One way to understand how this diagram shows that the eight events are equally likely is to think of throwing a dart randomly at the diagram. The dart has an equal chance of landing in any of the eight regions. Some students may reason more easily with tree diagrams, and other students may reason more easily with geometric regions.

Fig. **1.3.**
Geometric representations of probabilities

H	T

One coin, two outcomes: each probability is 1/2.

HH	TH
HT	TT

Two coins, four outcomes: each probability is 1/4.

HHH	THH
HHT	THT
HTH	TTH
HTT	TTT

Three coins, eight outcomes: each probability is 1/8.

Mathematically, tossing two coins has the same structure as counting the gender of babies at birth. Since the birth of a boy and the birth of a girl happen nearly the same number of times over the long term, the gender of a baby is often treated as one of two equally likely outcomes. Counting the number of boys in successive births is structurally like counting the number of heads in successive tosses of coins. Replacing *H* with *boy* and *T* with *girl* in the tree diagrams in figures 1.1 and 1.2 would yield the tree diagrams for the gender of two or three successive births, respectively. Since the tree diagrams are structurally the same, the underlying mathematical analyses would also be the same. This idea is important to helping students understand the notion of mathematical structure.

Tossing Two Dice

One of the most familiar situations in which the "obvious" outcomes are not equally likely is rolling two dice and summing the numbers on the top faces of the dice. The sums can range from 2 to 12, but the sums are certainly not all equally likely. The sum 7 is most likely to occur, and the probability decreases as the sums get progressively larger (or smaller). Sometimes students will say that "7 occurs most of the time." This statement is, of course, not true. The sum 7 is the most likely sum, but "sum different from 7" is much more likely than "sum 7." The probability of getting a sum equal to 7 is 6/36, or 1/6; 5/6 of the time the sum "not 7" can be expected to occur.

An important way to analyze the two-dice situation is to think of all the individual outcomes that might occur—that is, to examine the sample space, which consists of thirty-six simple events. Similar to thinking of two coins as different (e.g., a penny and a nickel), the two dice can be thought of as distinct—for example, a red die and a

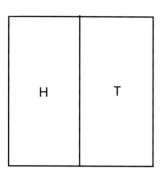

"Although the computation of probabilities can appear to be simple work with fractions, students must grapple with many conceptual challenges in order to understand probability."

(NCTM 2000, p. 254)

green die. There are thirty-six equally likely outcomes. With different-colored dice, it is easy to see, for example, that (3, 4)—meaning red 3 and green 4—is different from (4, 3)—meaning red 4 and green 3. Six of these combinations yield a sum of 7: (1, 6), (2, 5), (3, 4), (4, 3), (5, 2), and (6, 1), so the probability of a sum of 7 is 6/36, or 1/6. Listing all the combinations for any other sum is an important step toward determining the probability of that sum. Finding all the simple, equally likely events (e.g., determining that for the two-dice situation, thirty-six combinations are possible) is a common technique for analyzing complex sample spaces.

Comparing probabilities is done most directly by comparing rational numbers. Some students, however, will use the idea of odds; that is, a comparison of the number of favorable outcomes to the number of unfavorable outcomes. For example, the probability of rolling a seven is 6/36, or 1/6, but the odds of rolling a seven are 6:30, or 1:5. The odds show the ratio of favorable outcomes (6 out of 36) to unfavorable outcomes (30 out of 36). Over the long run, a sum of 7 will occur 6/30, or 1/5, as often as a sum other than 7. This strategy is less flexible, and it can result in confusion between odds and probabilities, but for some students, it is an alternative approach to probability problems.

One important context for helping students understand probability is the fairness of games. This topic may have been addressed in elementary school, but sometimes students do not internalize a mathematically appropriate notion of fairness. Fairness means that over many plays of a game, each player has the same chance of winning the game. It does not mean that students alternate winning the game or that each player actually wins the same number of games in, for example, a tournament. In a fair game, the probability of each player's winning over the long term is the same. If two students played great numbers of fair games, the relative frequency with which each won would begin to approximate one-half. Students must separate the notion of the probability of winning over the long term from the actual data for winning and losing in only a few plays of a game. They must also understand that this notion of fairness applies to particular games, such as rolling a die or spinning a spinner, in which the outcome does not depend on strategy or the skill of the player.

More-Advanced Concepts

Various other concepts are involved in examining sample spaces. *Mutually exclusive events* are events that cannot happen at the same time; that is, the events have no common elements in the sample space. For example, in rolling two dice, getting an odd sum and getting an even sum are mutually exclusive. Getting a sum divisible by 2 and getting a sum divisible by 3 are not mutually exclusive, however, since the sums 6 and 12 are each divisible by 2 and 3 simultaneously. In the more complex situations that students will encounter in grades 9–12, knowing when events are or are not mutually exclusive is often important, so teachers should begin to lay the groundwork in the middle grades.

Complementary events are mutually exclusive events that together encompass the entire sample space. For example "sum = 7" and "sum ≠ 7" are complementary events for a roll of two dice. For the toss of

The sums of the possible results of rolling two dice are represented in this matrix:

	\multicolumn{6}{c}{First Die}
Second Die	1	2	3	4	5	6
1	2	3	4	5	6	7
2	3	4	5	6	7	8
3	4	5	6	7	8	9
4	5	6	7	8	9	10
5	6	7	8	9	10	11
6	7	8	9	10	11	12

a coin, H and T are complementary events. The sum of the probabilities of all the outcomes in a sample space must be 1, since the sample space specifies exactly all the outcomes that can happen. Therefore, if we know the probability of an event, then the probability of "not that event" is simply the difference between 1 and the probability of the event.

Another important idea is *independent events.* Two events are independent if they do not influence each other; that is, the occurrence of one event does not influence the chance of occurrence of the other event. For example, rolling one number on the red die does not influence the outcome of rolling the green die. The two dice are independent of each other. In mathematics, two events are independent if the probability of the two events occurring simultaneously is the product of the probability of each event occurring separately. For example, if two dice are rolled, the list of outcomes when both dice give even numbers has nine elements: (2, 2), (2, 4), (2, 6), (4, 2), (4, 4), (4, 6), (6, 2), (6, 4), and (6, 6). The probability of two even numbers is 9/36, or 1/4. The list of the outcomes for an even number on one die has three elements: 2, 4, and 6. The probability of an even number on either die is 3/6, or 1/2. In this case, the probability of both dice showing even numbers is $1/2 \times 1/2 = 1/4$, the product of the probability of the red die showing an even number and the probability of the green die showing an even number. The events "red die showing even" and "green die showing even," then, are independent.

One of the instances in which an understanding of independence is important is replacement or nonreplacement in sampling. Suppose we start with a jar containing three blue candies and three yellow candies and that we take out one candy (without looking), eat it, and then draw a second candy. What is the probability that the second candy is yellow? Well, it depends on what color the first candy was. If the first candy was yellow, only two yellow candies are left (out of five), so the probability is 2/5; if, however, the first candy was blue, three yellow candies (out of five) are left, so the probability is 3/5. The probability of drawing a yellow candy on the second draw depends on the outcome of the first draw. If we draw without replacement, the events "yellow on first draw" and "yellow on second draw" are not independent but dependent. By contrast, if we had not eaten the first candy but had put it back in the jar, the probability of drawing a yellow candy on the second draw would still be 3/6, or 1/2. In other words, if replacement takes place, the events are independent.

What Might Students Already Know about These Ideas?

You can expect middle-grades students to demonstrate wide ranges of both understanding of probability concepts and skills in computing probabilities, especially for compound events. One way to assess their knowledge is by asking them to compare their intuitions about the likelihood of events with actual outcomes and to discuss their findings. Students are usually familiar with spinners used in games, so activities like the following, Who Will Win? (which involves spinners), can make a discussion of outcomes accessible to most students.

Who Will Win?

Goal

To assess students'—

- ability to identify which outcome on a spinner is most likely to occur;
- skill at using appropriate language to explain their reasoning.

Materials and Equipment

- A copy of the blackline master "Who Will Win?" for each student
- Paper clips, tape, pencils, protractors, compasses, paper, cardboard, and colored markers for creating spinners (optional)

p. 90

Activity

Set up the activity by asking the students what board games they have played, either at home or in school. The students are likely to mention a variety of games. Ask them what devices are used in those games to decide what moves should be made. Their responses will probably include dice and spinners; other methods, such as drawing the top card from a deck of cards, may also be mentioned.

Next, ask the students how they know when a game is fair. Do not try to teach the idea of fairness at this time; simply let the students get their ideas into the open. The discussion of the activity should help them develop a good understanding of fairness. Responses that you might expect from students include the following:

- "If each person wins the same amount"
- "If everyone has the same chance to win"
- "If you can't predict what will happen"
- "It's fair to me if I win more often."

Distribute an activity sheet to each student. The students should work individually on questions 1–4. You may want them to work together on questions 5 and 6 or to compare their answers with those of a partner before you discuss their work with the class as a whole.

Discussion

It is important that students be able to explain that any of the three colors might come up on each spin of the first spinner on the activity sheet. Investigating question 1 helps them develop the notions of outcomes and sample space. You can probe their thinking by asking, "Is each color equally likely to come up on each spin? How do you know?" Their comparisons of the likelihood of the occurrence of each of the three colors can help you evaluate their understanding. Almost all students will consider this spinner "unfair," since blue has such a big advantage on each spin. The probability of landing on blue is twice as great as the probability of landing on either other color. You should not expect all students to use the term *probability*, however.

Chapter 1: The Concept of Probability

Students' explanations might focus on the size of the angles whose rays delineate the three regions or on the area of each sector. Angle size is the better choice in this instance, since a game could be played with a spinner whose spin hole is off center, as shown in figure 1.4. Although the areas of the colored regions are not in the ratio 2:1:1, the probabilities are still 1/2, 1/4, and 1/4, since the measures of the angles centered on the hole are still in the ratio 2:1:1. Discussing such spinners may not be profitable at the beginning of probability instruction, but you might be alert for a teachable moment to expose your students to examples like the spinner in figure 1.4.

Fig. **1.4.**

A spinner with an off-center hole

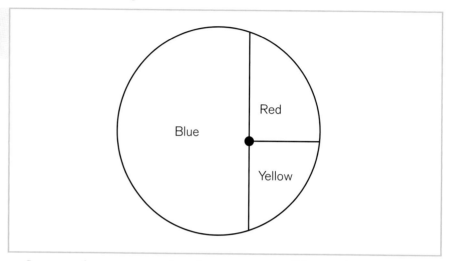

Some students may say that three outcomes (blue, red, yellow) are possible on each spin of the second spinner on the activity sheet, whereas others may say there are eight possible outcomes (blue, blue, blue, red, red, red, yellow, yellow). If the students don't suggest the latter of these responses, you might say that a student in another class suggested it and ask which way of listing the outcomes is correct. Of course, both are correct, but if only three outcomes are listed, they are not equally likely, whereas if all eight outcomes are listed (one for each sector), they are equally likely. The outcome blue is a compound event, consisting of landing on any one of the three blue sectors. The outcomes red and yellow are likewise compound events.

Most students will see this spinner as "unfair" also, but they may say that a game based on it is "more fair" or "closer to fair" than a game based on the first spinner. Discussing the second spinner presents an opportunity to talk about the "total for red" (i.e., the mental combining of the red sectors) to see if the students are thinking about summing the measures of the angles of the sectors or the areas of the regions. You might informally ask the students if they can tell you "how much off" from fair a game based on this spinner would be. The students may talk about needing "one more sector" of yellow, which could lead to a discussion of what size of sector would be needed, how it could be fit into the spinner, and what would happen to the existing sectors. In order to create nine congruent sectors, the size of each of the existing sectors would need to be reduced by making each angle smaller.

Students' solutions to questions 5 and 6 can reveal a lot about their thinking (about angle vs. area, for example). You could have several students present their spinners and then have the class discuss whether the

"Teachers should build a sense of community in middle-grades classrooms so students feel free to express their ideas honestly and openly, without fear of ridicule."
(NCTM 2000, p. 268)

spinners would produce fair games. Be alert for unusual solutions to this task.

Extension

The CD-ROM that accompanies this book includes two applets, Spinner and Adjustable Spinner, that allow students to create electronic spinners that can be used to explore some of the ideas of probability presented in this book. With Spinner, students can create spinners with one to twelve equal sectors of different colors. Adjustable Spinner allows students to design spinners with one to nine sectors of different colors and with varying angle measures

Alternatively, using a pencil, compass, protractor, paper, and colored markers, the students could design the faces of spinners and then follow these instructions: Cut out the spinners, and mount them on cardboard if necessary. Poke a small hole in the center of each spinner, using the point of a compass or a sturdy pen. Unbend the end of a paper clip and poke it through the hole from the back. Tape the rest of the paper clip to the back of the cardboard. Put a second paper clip on the protruding end of the first one to use as a needle. Make sure that the paper clip can spin freely. (See the illustration in the margin.)

Selected Instructional Activities

Although spinners are probably very familiar to students, dice or number cubes are somewhat easier to use in various settings, especially in simulations of events. In order to use dice effectively, however, students need to understand that when two dice are rolled, there are thirty-six equally likely outcomes. The next activity, Two Dice, will help students gain or review that knowledge.

The other activities in this chapter introduce students to the ideas of independent events, similarity of structure across situations, and probability as a relative frequency. These activities can serve as only an introduction to these important ideas. More experiences in a variety of settings are needed to help students develop a strong conceptual understanding that links the ideas of probability with other important mathematical ideas.

Two Dice

Goals

- Understand that rolling two dice once is equivalent to rolling one die twice
- Use a listing of equally likely outcomes to generate a sample space

Materials and Equipment

p. 91

- A copy of the blackline master "Two Dice" for each student or pair of students
- Dice or number cubes

Activity

The probability concepts involved in rolling two dice and summing the resulting numbers are important for students to understand. Students are likely to have different understandings of the situation, however. Some students may believe that the sums are equally likely or nearly equally likely.

Since rolling dice and summing the numbers on the top faces is a familiar activity, you can simply distribute the activity sheet and ask the students to begin. Most students should be able to do the work mentally, but have dice or number cubes available for those who need to manipulate them. The students can work individually or in pairs. As they work, watch for the different ways that they record the possible pairs of numbers. When they make lists, note how those lists are organized or whether they are organized appropriately. If the lists are not organized, you may want to help the students develop organizing strategies as part of the follow-up to the activity. In most classes, at least one student will make a tree diagram or use geometric representations, but if no one does, you may want to introduce those techniques during the follow-up.

Discussion

For questions 1 and 3, most students will be able to list the outcomes 2, 3, ..., 12. You may need to conduct a detailed discussion, though, about whether they are equally likely. You might ask, "How could you decide if they are equally likely?" The experimental, or empirical, approach is to toss the dice many times and see what happens. (If you have access to a computer simulation, you might use it.) Even a few dozen tosses should generate data that suggest that the sums do not come up equally often. So a deeper analysis of the situation is indicated.

This analysis can be approached through a discussion of the ways in which the students showed the pairs of numbers that could occur. Of course, the answers to questions 2 and 4 are the same: thirty-six pairs. It is much easier for students to see all the combinations if the dice are different colors. As the students discuss their answer to question 2, be sure that they are clear about which number in a pair is associated with the red die and which number is associated with the green die; "red 4

A *simulation* is a model of a problem that occurs in real life. The model is designed so that it reflects the probabilities of the real situation.

and green 3," for example, is different from "red 3 and green 4." Some students may use "pair notation"—that is, (3, 4) to show red 3 and green 4, for example—but this notation is not necessary. The goal is to list all thirty-six combinations clearly. The students may need some help in organizing their lists so that they are sure they have listed all possible pairs.

Once the lists are complete, ask, "Is each pair equally likely?" (yes) How do you know?" (Each number on each die is equally likely.) "How can the pairs be grouped to help you understand how likely each sum is?" (Pairs that generate the same sum could be grouped.) The process of grouping the pairs according to the sums that they generate may be new to some students. It is a crucial part of analyzing the situation, however, so it is important that students develop facility with it.

Rolling one die twice may be more familiar to some students and therefore easier for them to understand. There is no mistaking which number occurred first and which number occurred second; this method should make clear that, for example, 4 on the first roll and 3 on the second roll is different from 3 on the first roll and 4 on the second roll. The notion "first number and second number" is mathematically equivalent to the notion "number on the red die and number on the green die." The pairs of numbers possible in the first-die–second-die situation are the same as those in the red-die–green-die situation. Again, grouping pairs that generate the same sums will help the students see the similarity in structure of these two contexts. The students should be able to conclude that the sums for the two-dice situation have the same probability as the sums for the one-die-tossed-twice situation. This discussion should also result in students' understanding that when two dice are rolled—regardless of color—(3, 4) and (4, 3) must be considered different outcomes.

Question 6 will challenge students because it requires analyzing an unfamiliar situation. It is not necessary that students solve this problem completely on their own. An analysis can be conducted by examining the thirty-six possible pairs of numbers generated according to the instructions and grouping them according to the generated values. You may want to conduct this analysis with the class as a whole. The outcomes are given in table 1.1. The possible outcomes are 3, 4, ..., 18. The outcomes with a probability of 3/36, or 1/12, are 7, 8, 9, 10, 11, 12, 13, and 14. The outcomes with a probability of 2/36, or 1/18, are 5, 6, 15, and 16. The outcomes with a probability of 1/36 are 3, 4, 17, and 18. You could explore the students' understanding of this situation by posing questions such as "What is the probability of an outcome between 5 and 9, inclusive?" and "How do you know?"

The pairs (3, 1), (2, 2), and (1, 3) are grouped together, since they all generate the sum 4.

Table 1.1
The Outcomes of Multiplying the Results of the First Roll of a Die by 2 and Adding the Results of the Second Roll

Second Die	First Die					
	1	2	3	4	5	6
1	3	5	7	9	11	13
2	4	6	8	10	12	14
3	5	7	9	11	13	15
4	6	8	10	12	14	16
5	7	9	11	13	15	17
6	8	10	12	14	16	18

Chapter 1: The Concept of Probability

Students could also respond to question 6 using an experimental approach, which would involve tossing dice many times and tallying the number of times each outcome occurs. (A computer simulation could easily be written to generate large quantities of data, although students should actually roll dice several times to be sure that they understand the situation.)

Although this activity focuses on the careful, accurate, and thorough analysis of situations, some attention should also be given to the language that students use in explaining their analyses. The next activity, More Often and Most Often, places a greater emphasis on the language that is used to report observations, conclusions, and generalizations. Intuitions about probability are quite strong—even though they are often erroneous—and the language mathematicians use to talk about probability often has other, common meanings that can lead to confusion. It is important for students to understand how differences in language convey differences in meaning.

More Often and Most Often

Goals

- Use precise mathematical terminology to describe probability situations
- Correctly interpret and reason about probability situations
- Explain thinking accurately

Materials and Equipment

- A transparency copy of the blackline master "Pets"
- A copy of the blackline master "More Often and Most Often" for each student
- An overhead projector

pp. 93, 94

Activity

Introduce the activity by displaying on the overhead projector the graph "Pet Owners Who Own Only One Pet" (on the blackline master "Pets"), which shows the number of people in a sample of pet owners who own different kinds of pets. Say, "A student who was looking at this graph said that most people had cats as pets. Is that true?" Many students will agree with this claim. In common language, the mode is often described by saying, "Most of the time it is this." Such a statement is often erroneous and is not correct in this case. There are more pet owners who do not have cats than pet owners who do have them; that is, "not cats" is more frequent than "cats" among the pet owners in this sample. A more accurate description is "More pet owners with only one pet have cats than any other pet" or "The most commonly owned type of pet is a cat." If some students disagree with the student's statement, ask them to explain why. If they do not disagree, ask, "Are more people in this sample cat owners or not cat owners?" "Would it be correct to say that most pet owners do not own cats?" (Most pet owners in this sample are non–cat owners. It is not true that most of the people own cats.)

Part of the reason for misstating the meaning of the mode is the multiple interpretations of *most*. Technically, the word means *more than one-half*, although in common parlance, it is sometimes used to indicate *most frequent* or *more than any other choice*. If the latter definition were used, the statement "Most people have cats as pets" would be true for the sample. Since precise language is so important in mathematical discourse, this activity is designed to challenge students' typical use of language related to *more* and *most* to help them become both more precise themselves and more sensitive to precision (or imprecision) in others' language.

Students should work individually on the activity sheet "More Often and Most Often." It may be beneficial for them to compare their answers with those of a partner before you discuss the activity with the class as a whole.

Discussion

This activity will help students refine their use of language. In question 1, for example, it is common for students to say that mint is likely to be drawn most of the time, when they really mean that mint is the flavor most likely to be drawn. Many students find this difference subtle, but it is important, and they must refine the precision of their language to reflect the difference. That something will happen most of the time means that it will happen more often than it will not happen. That something is most likely to happen means that it will happen more often than any other single option (rather than the combination of all other outcomes). This difference is important mathematically.

The differences in the language in the questions are intentional. Specifically, the use of *probability*, *more likely*, and *likely most of the time* gives you and your students an opportunity to talk about whether the meanings of these phrases are the same or different. The essential differences among these terms are that some are appropriate to describe what happens most of the time but others are appropriate to describe what happens more often than any other single option. In the discussion following this activity, it is important to attend carefully to the language that students use to explain their answers. The following questions may be helpful in guiding the discussion:

- What outcomes are possible each time a candy is drawn? (mint, with probability 1/2; butterscotch, with probability 1/3; and caramel, with probability 1/6)
- Are the outcomes mint, butterscotch, or caramel equally likely? (No. You could, however, mention six equally likely outcomes: mint, mint, mint, butterscotch, butterscotch, caramel.)
- Which outcome is most likely? (mint) How do you know? (There are more mints than candy of any other flavor.)
- Which outcome is least likely? (caramel) How do you know? (There are fewer caramels than candy of any other flavor.)
- If you had to predict a single outcome, what flavor would you pick? (mint) Why? (It has the greatest probability of being drawn.)
- Suppose that you drew out a candy one hundred times, replacing the candy each time. Describe an event that might have occurred most of the time. Remember that *most of the time* means *more than half the time*. (Two possible answers are (1) not butterscotch [which is the same as "mint or caramel"] and (2) "mint or butterscotch.")
- Is it correct to say, "Mint would be drawn most of the time"? If not, how could you change the contents of the bag so that this statement would be true? (Mint can be expected to be drawn about 50 percent of the time, which is not most of the time. If one or more mint candies were added to the bag or if one of the nonmint candies were removed, then mint would be expected to be drawn most of the time from the modified bag.)
- What are the similarities and differences among the statements "Mint is drawn more often than any other flavor," "Mint is drawn most often," and "Mint is drawn more than the other flavors"? [Some students may suggest that the differences in language highlight the difference between most of the time (more than 50%) and most frequent (most likely).]

This task will help students develop their mathematical communication skills so that their statements will be more precise and accurate, even when the mathematical meaning of terms may be different from the common, often erroneous meanings of the same terms in everyday conversation. Mathematical communication is an important component of mathematical understanding.

Extension

If your students know about permutations and combinations, you can present the following statements about the candy situation and ask the students if they are true or not true. Ask the students also to explain their answers.

- If you drew out two candies at once, they would both be mints most of the time.
- If you drew out two candies at once, they would both be the same flavor more often than they would be different flavors.

Remember that the mint candies are distinct, so we can think of them as M_1, M_2, and M_3, and the butterscotch candies are distinct, so we can think of them as B_1 and B_2. Since the bag contains six candies, the following fifteen pairs can be drawn out (M means mint, B means butterscotch, and C means caramel):

M_1M_2, M_1M_3, M_1B_1, M_1B_2, M_1C
M_2M_3, M_2B_1, M_2B_2, M_2C
M_3B_1, M_3B_2, M_3C
B_1B_2, B_1C
B_2C

Three of the pairs—M_1M_2, M_1M_3, and M_2M_3—consist of two mints, so the probability of drawing out two mints is 3/15. Four of the pairs—M_1M_2, M_1M_3, M_2M_3, and B_1B_2—consist of candies of the same kind, so the probability of drawing out two candies of the same flavor is 4/15. Both of the statements, then, are not true.

The next activity, Empirical Probabilities, returns students to analyzing rolls of the dice. In it, students are introduced informally to empirical probabilities, a topic that is treated in more detail later in the book. Their mathematical language continues to be expanded, although in a slightly different way. In Empirical Probabilities, students identify patterns in data and describe them as ratios of positive outcomes to total numbers of outcomes. Students are also encouraged to think about the sample space of sums on two dice—for example, "prime" and "not prime," which are complementary events.

Communicate ... mathematical thinking coherently and clearly to peers, teachers, and others

Empirical Probabilities

Goals

- Understand relative frequency as empirical probability
- Explore how empirical probability stabilizes as the number of rolls of the dice increases

Materials and Equipment

p. 95

- A copy of the blackline master "Ratios" for each pair of students
- Dice
- Calculators

Activity

Since rolling dice is familiar to most students, the introduction to this activity can be as simple as reviewing the possible sums of the values on the faces of two dice: 2, 3, 4, ..., 12 and reviewing the difference between prime and nonprime numbers. You might also ask the students to list the possible products of the values on the faces of two dice. The least product is 1, and the greatest product is 36, but some numbers between 1 and 36 (e.g., 11 and 23) cannot be generated by multiplying the values on the faces of two dice. Allow the students a few minutes to complete a list of the products.

Distribute the activity sheets, and have the students work in pairs. Some students may need assistance in understanding the process of finding the ratio of "number of prime sums" to "cumulative number of rolls so far." The important idea is that "cumulative number of rolls so far" keeps changing; it increases by 1 with each new roll. The vertical axes for the graphs on the activity sheet are marked in decimals, so it is important that the ratios be written as decimal fractions. The students will probably find it useful to have a calculator available for doing computations, and they may also benefit from comparing their answers with those of another pair before you discuss the activity with the whole class.

Discussion

One of the important ideas underlying this activity is that the ratios generated are *empirical probabilities;* that is, they are generated from trials of an experiment. The settings can be analyzed theoretically also, so the students can observe whether their experimental probabilities approach the *theoretical probabilities.*

Another idea is that in most cases, the variability in the ratios for the first ten rolls will be greater than the variability in the cumulative ratios as the number of rolls increases. As more data are gathered, the empirical probability can be expected to approach the theoretical probability. This idea is addressed in chapter 3. For both situations (the sum and the product), the charts should demonstrate that the cumulative ratios for large numbers of rolls are closer together than the ratios for the first ten rolls.

The prime sums are 2, 3, 5, 7, and 11. The sum of the probabilities of these sums is 15/36, or about .42. The data generated by any pair of students may not seem to be very close to this value, but if you pool the data for the class, the pooled estimate should be much closer. This activity affords an opportunity to discuss the difference between each pair's data and the data for the class as a whole. In response to question 4, some students may describe the pattern they see in their table; that is, they may base their answer on empirical probabilities. Other students may base their answers on the thirty-six possible outcomes for rolling two dice; that is, they may answer on the basis of the theoretical probabilities. Challenge the students to compare the differences in these kinds of reasoning, and lead a discussion about which type of reasoning is more believable. The students are likely not to agree on this issue, which is acceptable at this point in their discussion of probability.

The possible products of the two numbers on two dice are 1, 2, 3, 4, 5, 6, 8, 9, 10, 12, 15, 16, 18, 20, 24, 25, 30, and 36. Table 1.2 displays this information. The thirty-six ordered pairs can be grouped on the basis of equal products. The probability of each of the products is small, ranging from 1/36 to 4/36 (see table 1.3). Of the thirty-six pairs of numbers, nineteen generate two-digit products, so the probability of generating a two-digit product is 19/36, or about .53. In response to question 8, the students are again likely to argue both the empirical-probability point of view and the theoretical-probability point of view. Since the theoretical reasoning involved in this problem is more complicated than that for the sums-of-numbers problem, you can expect many students to rely on the empirical approach.

Table 1.2
Products of the Possible Values on the Top Faces of Two Dice

Green Die	Red Die					
	1	2	3	4	5	6
1	1	2	3	4	5	6
2	2	4	6	8	10	12
3	3	6	9	12	15	18
4	4	8	12	16	20	24
5	5	10	15	20	25	30
6	6	12	18	24	30	36

Table 1.3
Probabilities of the Products of the Possible Values on the Top Faces of Two Dice

Products	Frequency	Probabilities
1	1	1/36
2	2	2/36
3	2	2/36
4	3	3/36
5	2	2/36
6	4	4/36
8	2	2/36
9	1	1/36
10	2	2/36
12	4	4/36
15	2	2/36
16	1	1/36
18	2	2/36
20	2	2/36
24	2	2/36
25	1	1/36
30	2	2/36
36	1	1/36

The graphs of both sets of data may be somewhat challenging for students to make and analyze. First, they have to understand the notion of ratio, which is not an easy concept for students. Second, they have to understand how to graph points like (6, 0.67). Some students may not be used to graphing points whose coordinates are not integers. Third, there is likely to be more variation in the ratios for the first ten rolls than for a larger number of rolls, which may present some difficulties in analyzing the graphs.

Conclusion

The analyses required for the activities in this chapter lay the groundwork for students' understanding of important probability concepts. The next chapter makes more explicit the notion of a *probability distribution*, which is the analog of a data distribution, discussed in *Navigating through Data Analysis in Grades 6–8* (Bright et al. 2003). Through formalizing their intuitive ideas about probability, students can gain the tools they need for more-sophisticated analyses.

Navigating through Probability

Grades 6–8

Chapter 2
Probability Distributions

Probability theory is the mathematical way of modeling the likelihoods of outcomes in random events.

The distributions discussed in this chapter are *discrete distributions,* since there are only a finite, or countable, number of outcomes, such as the six outcomes of rolling a die or the thirty-six outcomes of rolling two dice. Distributions that describe continuous variables, such as all the real numbers in an interval on the number line, are beyond the scope of the middle-grades curriculum.

Important Mathematical Ideas

Probability is a way to think about random events or random experiments. Random events lack any definite or predictable pattern of outcomes. For example, when we spin a spinner with regions of different colors, we know all the outcomes that could happen each time we observe the event—that is, all the colors the spinner might land on—but we do not know which outcome will happen on any particular observation. Rolling dice and tossing coins are two random events that can be thoroughly analyzed. Other, more complex examples include gender in litters of animals, the length for life of light bulbs, or the number of correct answers generated by guessing all the answers on a multiple-choice test.

Probability Distribution

Each random event has an associated *probability distribution*, which is a specification of all the possible outcomes of that event, along with the corresponding theoretical probabilities for those outcomes. Two mathematical requirements must be satisfied by a probability distribution:

1. The probability assigned to each outcome must be a number between 0 and 1, inclusive.
2. The sum of all the probabilities assigned must be 1.

The probability distribution of a random event is a complete mathematical description of that event; it contains all the relevant probabilistic information about the event.

A probability distribution is usually represented in one of three ways: by a table, by a graph, or by a formula. The spinner in figure 2.1, for example, is represented by the table and graph in figure 2.2. A table of a probability distribution shows all the outcomes and the probabilities associated with each of those outcomes. A graph of a discrete probability distribution looks like a bar graph. The outcomes are displayed on the horizontal axis, and the probabilities are shown by rectangles of equal width above those values; the heights of the rectangles are proportional to the probability of the associated outcomes. When the outcomes are numerical, it is sometimes possible to give a formula for calculating the probability of each outcome. Formulas are frequently used by mathematicians as a way of describing probability distributions, but the level of mathematical sophistication needed to understand most of the formulas goes beyond the normal middle-grades curriculum. Algebra students sometimes use the binomial expansion—the expansion of $(x + y)^n$—to show the outcomes for a binomial random event—n independent trials of an event with two possible outcomes on each trial, such as tossing a coin n times.

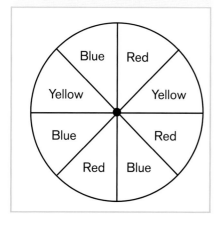

Fig. **2.1.**

A spinner with eight equal sectors distributed among three colors

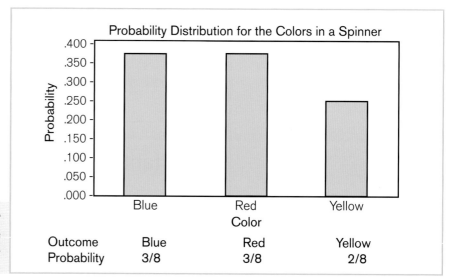

Fig. **2.2.**

A graph and a table representing the probability distribution for the spinner in figure 2.1

We can create a probability distribution for the spinner in figure 2.1. The outcomes are blue, red, and yellow, which can be expressed in set notation: {blue, red, yellow}. It is reasonable to assign a probability to each color on the basis of the number of congruent sectors of each color (i.e., on the basis of the relative sizes of the angles associated with the colors). The table and graph in figure 2.2 show these outcomes and their probabilities; since the outcomes are not numeric, there is no formula for showing this probability distribution.

Probabilities could be assigned in many ways that would satisfy the two rules on page 29. The choice depends on the particular experiment to be modeled. For instance, if a coin is tossed twice, the outcomes are HH, HT, TH, and TT. The probability of each outcome can therefore be determined to be 1/4. However, suppose that the coin is biased (i.e., weighted) so that a head is twice as likely as a tail on each toss. Then a different assignment would be more appropriate, since certainly HH should be more likely than TT. In this case, $P(H) = 2/3$ and $P(T) = 1/3$ on each toss. The coin tosses are independent, so the probability of observing two heads is the product of the probability of heads on each coin. That is, $P(HH) = P(H) \cdot P(H) = 2/3 \cdot 2/3 = 4/9$. Similarly, we can compute the theoretical probabilities of each of the other outcomes: $P(HT) = P(TH) = 2/9$, and $P(TT) = 1/9$. Each of these probabilities is between 0 and 1, and $4/9 + 2/9 + 2/9 + 1/9 = 1$; therefore, this assignment of probabilities to the four outcomes fits the two mathematical requirements.

Describing the shape of a probability distribution is an important part of understanding the distribution. The shape might reveal information about which outcomes are more likely or less likely to occur. For example, figure 2.3 shows the probability distributions for the number of heads on tosses of one coin and three coins. It is very clear from the shape of the displays that the two outcomes for the one-coin situation are equally likely and that the four outcomes for the three-coin situation are not equally likely. Observing one or two heads for three coins is more likely than observing zero or three heads.

Independent events are discussed in chapter 1.

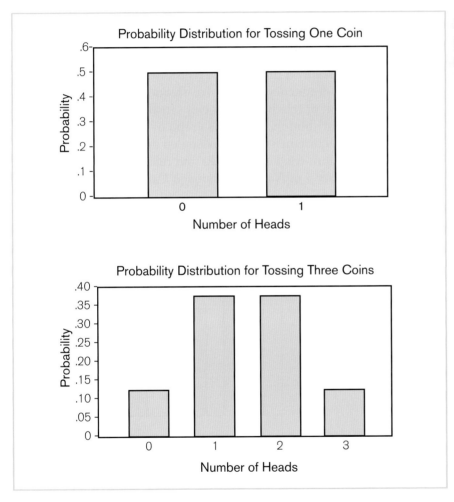

Fig. **2.3.**

Probability distributions for tossing one coin and three coins

The assignment of probabilities to the outcomes of a particular random event are influenced by many factors. Physical aspects, such as the geometry of spinners or the balance of coins, are often given the highest consideration. Techniques for mathematical analysis, such as tree diagrams or geometric probability diagrams, are also useful tools in creating a probability distribution.

Sometimes an empirical approach can be used to create a representation that is similar to the probability distribution. This experimental process involves making many observations of the event, recording the number of times each outcome occurs, and calculating the relative frequency of each outcome. The relative frequencies are called *empirical probabilities*. A table or a graph of relative frequencies displays the distribution of empirical probabilities and will often reveal patterns in outcomes. (Table 2.1 shows data for the spinner in fig. 2.1.) Such a display is often similar to the theoretical probability distribution for the random event. The difference is that the probability distribution shows the theoretical probabilities and the graph of relative frequencies shows the empirical probabilities. Only rarely are the empirical probabilities exactly the same as the theoretical probabilities, but as the number of observations increases to a very large number, we can expect the empirical values to approach the theoretical values. We can not be sure, however, that the empirical values will ever exactly equal the theoretical values. (This idea is discussed in more detail in chapter 3.)

A probability distribution or a graph of empirical probabilities is different from a bar graph of counts, which is a common way to represent tallies of data. A bar graph shows the absolute frequency of each outcome. It has the same shape as a graph of the relative frequencies of the outcomes, but the scale on the vertical axis is different. Figure 2.4 shows the absolute-frequency and relative-frequency graphs of the data in table 2.1. A graph of relative frequencies satisfies both requirements for a probability distribution, whereas a graph of absolute counts does not satisfy either requirement. A more detailed discussion of the difference between absolute-frequency graphs and relative-frequency graphs is available in chapter 3 of *Navigating through Data Analysis in Grades 6–8* (Bright et al. 2003).

> "Teachers can help students relate probability to their work with data analysis and to proportionality as they reason from relative-frequency histograms."
> (NCTM 2000, p. 254)

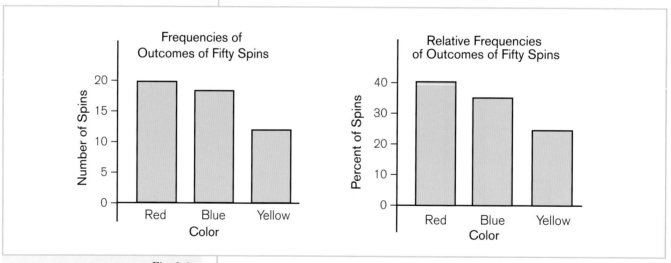

Fig. 2.4.

Absolute-frequency and relative-frequency graphs of the data in table 2.1

A probability distribution is an arbitrary notion; it may not seem "natural" to some students. For example, the mathematical requirements for a probability distribution would not be violated by saying that the probability of each of the three colors on the spinner in figure 2.1 is 1/3. However, our sense of correctness would certainly be violated. The sums of the central angles associated with the sectors of the same color are not equal, which we would expect if the probabilities were equal. It is often necessary to analyze the physical aspects of a situation in order to determine a correct probability distribution.

Table 2.1

Relative Frequencies of Outcomes of Spinning a Spinner with Eight Sectors Distributed among Three Colors in the Ratio 3:3:2

Spin	Outcome	Relative Frequencies of Outcomes		
		Red	Blue	Yellow
1	Blue	.00	1.0	.00
2	Blue	.00	1.0	.00
3	Yellow	.00	.67	.33
4	Yellow	.00	.50	.50
5	Red	.20	.40	.40
6	Red	.33	.33	.33
7	Blue	.29	.43	.29
8	Blue	.25	.50	.25
9	Yellow	.22	.44	.30
10	Red	.30	.40	.30
11	Blue	.27	.45	.27
12	Blue	.25	.50	.25
13	Red	.31	.46	.23
14	Blue	.29	.50	.21
15	Yellow	.27	.47	.27
16	Yellow	.25	.44	.31
17	Red	.29	.41	.29
18	Blue	.28	.44	.28
19	Blue	.26	.47	.26
20	Yellow	.25	.45	.30
21	Red	.29	.43	.29
22	Blue	.27	.45	.27
23	Red	.30	.43	.26
24	Yellow	.29	.42	.29
25	Yellow	.28	.40	.32
26	Blue	.27	.42	.31
27	Blue	.26	.44	.30
28	Red	.29	.43	.29
29	Red	.31	.41	.28
30	Red	.33	.40	.27
31	Blue	.32	.42	.26
32	Red	.34	.41	.25
33	Yellow	.33	.39	.27
34	Red	.35	.38	.26
35	Yellow	.34	.37	.29
36	Blue	.33	.39	.28
37	Red	.35	.38	.27
38	Red	.37	.37	.26
39	Blue	.36	.38	.26
40	Red	.38	.38	.25
41	Blue	.37	.39	.24
42	Blue	.36	.40	.24
43	Red	.37	.40	.23
44	Red	.39	.39	.23
45	Yellow	.38	.38	.24
46	Yellow	.37	.37	.26
47	Blue	.36	.38	.26
48	Red	.38	.38	.25
49	Red	.39	.37	.24
50	Red	.40	.36	.24

Random Variables

A random event is often best analyzed by creating a random variable for the event. The random variable allows us to assign a number to each element of the sample space. For example, for the tossing of two fair coins, we can create a random variable by counting the number of heads for each toss. The values of this random variable are 0, 1, and 2. As before, the adjective *random* refers to the fact that we do not know what the value of the variable will be for any particular observation. That is, each time we toss two coins, the number of heads is unpredictable, even though we know ahead of time that the number of heads will be either 0, 1, or 2. The probability distribution for the event "number of heads for a toss of two coins" shows the relative probabilities for each of the three outcomes:

Number of Heads	0	1	2
Probability	1/4	2/4	1/4

> A *random variable* is a function from the sample space to the real numbers. That is, the input for the function is an element of the sample space, and the output of the function is a real number.

If the coins are not fair—that is, if they are unbalanced—the probability distribution will change. Table 2.2 shows different probability distributions for coins with various degrees of bias—that is, for different values of $p = P(H)$. Figure 2.5 displays graphical representations of the probability distributions for the same coins.

Table 2.2
Probability Distributions for Tosses of Two Coins with Various Degrees of Bias

Value of p	$P(TT)$	$P(HT)$	$P(HH)$
1/4	9/16	6/16	1/16
1/3	4/9	4/9	1/9
1/2 (fair coins)	1/4	2/4	1/4
2/3	1/9	4/9	4/9
7/8	1/64	14/64	49/64

> As an example, consider tossing two unfair (biased) coins, each having $P(H) = 1/4$:
>
> $P(HH) = P(H) \cdot P(H)$
> $\Rightarrow 1/4 \cdot 1/4 = 1/16$
> $P(HT) = P(H) \cdot P(T)$
> $\Rightarrow 1/4 \cdot 3/4 = 3/16$
> $P(TH) = P(T) \cdot P(H)$
> $\Rightarrow 3/4 \cdot 1/4 = 3/16$ $\Rightarrow \frac{6}{16}$
> $P(TT) = P(T) \cdot P(T)$
> $\Rightarrow 3/4 \cdot 3/4 = 9/16$

The probability distribution when $p = 1/2$ is symmetric, whereas the probability distributions for other values of p display skewness. In skewed distributions, the data are bunched at one end of the range and stretched out at the other end. Skewed distributions are named according to the direction of the tail, not the location of the bulge, so we say, for instance, that a J-shaped curve is skewed left. More information on skewness can be found in chapter 2 of *Navigating through Data Analysis in Grades 6–8* (Bright et al. 2003). Observing the shape of a distribution helps in the analysis of likelihood. For example, a greater number of heads in n tosses is more likely when p is greater for each individual toss.

Many different random variables can often be defined for a particular sample space. For example, in the two-coin situation, we could define a random variable as the product of the number of heads and the number of tails. The values of this random variable are 0 (for 0 heads and 2 tails or 2 heads and 0 tails) and 1 (for 1 head and 1 tail), and the associated probabilities are 2/4 and 2/4, respectively. We could also define a random variable as the sum of the number of heads and the number of tails, but this variable is not very interesting. Its only value is 2, and the associated probability is 1.

> If p is the probability of a head for one coin, the four outcomes HH, HT, TH, and TT have probabilities of p^2, $p(1-p)$, $(1-p)p$, and $(1-p)^2$, respectively, since the two coins are independent. Also, $P(1 \text{ head}) = P(HT \text{ or } TH) = P(HT) + P(TH)$, since the outcomes HT and TH are mutually exclusive.

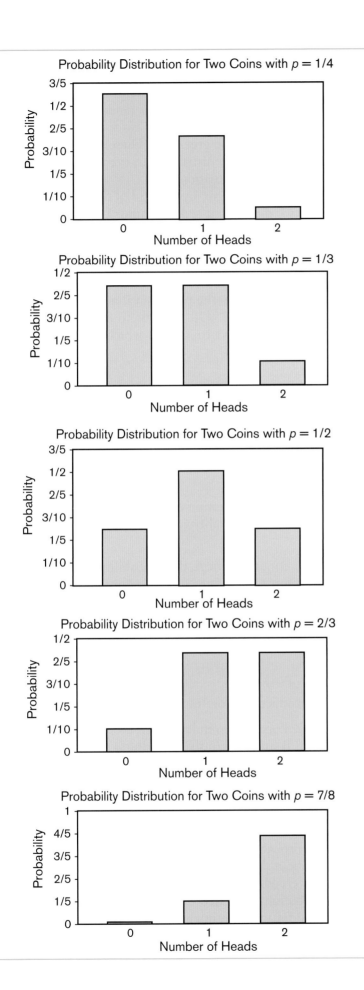

Fig. 2.5.

Probability distributions for tosses of two coins with various degrees of bias

Note that $p = P(H)$; that is, p is the probability of heads on one flip of a coin.

Wiest and Quinn (1999) explain games through which students can explore defining random variables.

More-interesting random variables can be defined for rolling two dice. Random variables that come immediately to mind are the sum of the two numbers, the difference between the two numbers, and the product of the two numbers. A less obvious random variable for tossing a red die and a green die is to compute twice the number on the red die plus the number on the green die. These random variables have very different probability distributions, even though they are generated by the same physical situation.

It is not difficult to use the six-by-six grid of the equally likely outcomes of rolling two dice (discussed in chapter 1; see table 1.2), determine the value of the random variable for each of the outcomes, count the number of outcomes associated with each value of the random variable to determine the theoretical probability of each value, and then make a probability distribution. For example, for the product of the two numbers on two dice, the possible values are 1, 2, 3, 4, 5, 6, 8, 9, 10, 12, 15, 16, 18, 20, 24, 25, 30, and 36. To find the probability that the product is 12, we have to identify which pairs generate a product of 12:

$$P(\text{product is } 12) = P(2, 6) + P(3, 4) + P(4, 3) + P(6, 2)$$
$$= \frac{1}{36} + \frac{1}{36} + \frac{1}{36} + \frac{1}{36}$$
$$= \frac{4}{36}$$

Once the probability distribution of a random variable is known, the following three rules of probability are helpful in calculating other probabilities:

1. *The "not" rule:* For any event A, the probability that A does not occur (often referred to as "not A") is 1 minus the probability of A; that is, $P(\text{not } A) = 1 - P(A)$.

   Example: $P(\text{product is not } 12) = 1 - 4/36 = 32/36$

2. *The "or" rule:* For any two mutually exclusive events A and B, the probability of A or B occurring is the sum of their probabilities; that is, $P(A \text{ or } B) = P(A) + P(B)$.

   Example: $P(\text{product is } 12 \text{ or } 36) = 4/36 + 1/36 = 5/36$

3. *The "and" rule:* For any two independent events A and B, the probability of A and B simultaneously occurring is the product of their probabilities; that is, $P(A \text{ and } B) = P(A) \cdot P(B)$. This rule is exactly the mathematical definition of independence given in chapter 1.

All the probabilistic information about a random experiment is carried through the probability distribution of the associated random variable. Questions about the likelihood of events arising from the experiment can be answered by the use of the "and," "or," and "not" rules, the probability distribution associated with the experiment, and the random variable described for the experiment. The activities in this chapter help illustrate these ideas.

When two events are not mutually exclusive, the "or" rule can be modified: $P(A \text{ or } B) = P(A) + P(B) - P(A \text{ and } B)$. This modification may go beyond most middle-grades curricula.

What Might Students Already Know about These Ideas?

Students are introduced to chance in the lower grades through various hands-on activities. The following activity, Fair Spinners, helps assess students' understanding of relative frequencies as estimates of probabilities and helps assess their use of data to support conclusions. The context is a setting that, although not realistic, should make sense to most students.

Fair Spinners

Goal

To assess students'—

- computation of relative frequencies;
- comparison of absolute-frequency and relative-frequency graphs;
- skill at using data to support conclusions about the fairness of spinners.

Materials and Equipment

- A copy of the blackline master "Fair Spinners" for each student
- Calculators (optional)

p. 99

Activity

The context of this activity is fictitious but is related to real-world situations. Introduce the activity by asking students what they know about how manufacturers assess the quality of their products. One common process is to choose a few products at random and test those products to be sure that they meet all specifications. The context in this activity is a simplified version of that process.

Ask the students what might make a spinner unfair. Some possible answers are that the cardboard is warped or torn or the spinner does not spin freely. You may also want to review briefly how to compute relative frequencies for data.

Distribute the activity sheets to each student to complete individually. The students are likely to be more successful at this task if they are permitted to use calculators.

Discussion

Part of the complexity of this activity is drawing double-bar graphs, one for each of the absolute frequencies and one for each of the relative frequencies. If the students have difficulty with this part of the activity, review the process for constructing these graphs during the follow-up discussion.

The most important aspect of this activity is the comparison of the two types of graphs. For a single set of data, the absolute-frequency graph and the relative-frequency graph will have the same shape. But when two sets of data with different Ns are graphed together, the comparison is more complicated, since the relative sizes of the double bars may change from one graph to the other. (See figs. 2.6 and 2.7.) In this case, the differences between the bars in the pairs in the absolute-frequency graph are visually greater than the differences between the bars in the pairs in the relative-frequency graph. Also, in the absolute-frequency graph, the bars for Pat's data are all taller than the corresponding bars for Chris's data because Pat made more spins. In the relative-frequency graph, which is a more appropriate representation for determining the fairness of the spinners, the pattern of taller bars is different. When asked to decide whether the spinners are fair, the

---

The relative frequency of an outcome is the decimal representation of the number of times that outcome occurred divided by the number of trials. For example, if a die is rolled ten times and "2" occurs three times, the relative frequency of "2" is .30 (3/10 = .30).

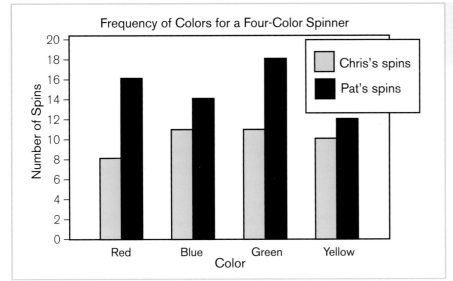

Fig. **2.6.**

An absolute-frequency graph for Chris's and Pat's spinner data

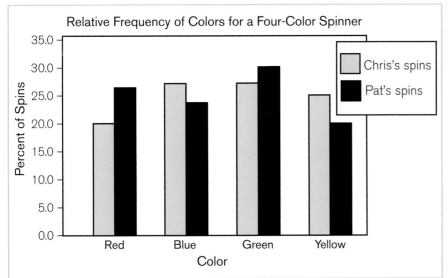

Fig. **2.7.**

A relative-frequency graph for Chris's and Pat's spinner data

students may focus their attention on a comparison of Pat's and Chris's data rather than the number of times (or how frequently) each of the four colors appears in each set of data. In each set of data, the relative frequencies of the four colors are close to the expected frequencies of 25 percent for each color. Pat's and Chris's data do not, then, suggest that either spinner is unfair.

Some students will notice that Pat's bars in the absolute-frequency double-bar graph (fig. 2.6) are visually more discrepant than Chris's bars, and they may reason that Chris's spinner is more nearly fair than Pat's spinner. Other students may focus on the visual differences in Pat's and Chris's bars for each of the colors in the absolute-frequency graph; for example, the black bars seem quite a bit different from the green bars. Students who reason these ways seem to be ignoring "the big picture" of the data in each set.

Other students will understand that relative frequencies are a better way of thinking about this situation. That is, since the bars on the relative-frequency double-bar graph are all close to 25 percent (see fig. 2.7), there is no reason to suspect that the spinners are not fair. Students who reason this way would seem to be ready to understand the law of large numbers—that is, it is important to focus on long-term trends and on relative frequencies in large data sets.

Selected Instructional Activities

The activities in this chapter will probably not repeat those that students completed in elementary school. One of the goals in each activity is to create and examine either a probability distribution or a relative-frequency distribution that approximates a probability distribution. Discussing each distribution as a whole, rather than focusing just on one or two outcomes for a distribution, helps develop an understanding of what a probability distribution is.

Your choices of which of the activities to use should be determined by your knowledge of your students' backgrounds. Collectively, the activities acquaint students with the probability distributions for tossing one die, two dice, and one coin.

The next activity, Dice Differences, uses a familiar task (i.e., rolling two dice) but asks students to examine the differences of the two numbers that were rolled rather than the sum of the two numbers.

"Teachers should give middle-grades students numerous opportunities to engage in probabilistic thinking about simple situations from which students can develop notions of chance."
(NCTM 2000, p. 253)

Dice Differences

Goals

- Understand sample space and the numeric outcomes associated with it
- Organize the results of a random experiment
- Understand the pooling of results
- Calculate the relative frequencies of outcomes and interpret them as probabilities

Materials and Equipment

- A copy of the blackline master "Dice Differences" for each pair of students
- A pair of dice for each pair of students

p. 102

This game was used by Freda (1998) to introduce probability.

Activity

The students should play the game in pairs. Each pair should assign one student to be player A and the other to be player B. Each player rolls a pair of dice. If the numbers indicated on the top faces are different, the player subtracts the lesser value from the greater. For example, if the numbers are 6 and 3, the difference is 6 – 3 = 3. If the numbers are the same, the difference is zero. For example, if the numbers are 4 and 4, the difference is 4 – 4 = 0. Before play begins, be sure the students understand that the difference will be either 0 or positive. If they are familiar with the term *absolute difference*, you can explain that they are to compute the absolute difference of the numbers on the two dice.

Distribute the activity sheets to each pair of students, and go over the rules, which are explained on the first page. On each roll, player A scores a point if the difference is 0, 1, or 2, and player B scores a point if the difference is 3, 4, or 5. You may want to play one or two demonstration rounds. Have the students record the outcomes of their games and tally their points on the activity sheet. Then have them record on the chart the number of times each difference occurred. Each game is played until one player has accumulated 10 points.

Discussion

After each pair of students has completed playing at least three games, bring the class together. Pool the data on the number of games won by players A and B. Ask the students if the game seems fair. They will probably agree that it is not a fair game. Ask them how these data show that the game is not fair. They will probably agree that since A and B do not win about equally often, the game is not fair. Some students may start to make an argument based on the probabilities of players A and B scoring points, but try to put off a consideration of the theoretical probabilities at this time.

Pool the data about the number of times each difference occurred. Compute the relative frequency of each difference. For the pooled data, the relative frequencies should be close to the theoretical probabilities, though the data for any particular game may not be very close. The theoretical probabilities can be computed from the information in table 2.3:

Difference	0	1	2	3	4	5
Probability	6/36	10/36	8/36	6/36	4/36	2/36
Decimal approximation	.17	.28	.22	.17	.11	.06

Table 2.3
Possible Outcomes of the Absolute Difference between the Values Resulting from Rolls of Two Dice

Second Die	First Die					
	1	2	3	4	5	6
1	0	1	2	3	4	5
2	1	0	1	2	3	4
3	2	1	0	1	2	3
4	3	2	1	0	1	2
5	4	3	2	1	0	1
6	5	4	3	2	1	0

Ask the students to add the probabilities of a difference of 0, 1, or 2 (for which player A scores a point) and the probabilities of a difference of 3, 4, or 5 (for which player B scores a point). The probability that A scores a point (24/36) is twice as great as the probability that B scores a point (12/36). Then ask the students how these probabilities are related to the fact that A won more games. Help the students see that since A is twice as likely as B is to score a point on each play, A is much more likely to accumulate 10 points before B does. It is quite unlikely that B would win any game consisting of ten rounds, even though B would be likely to score some points.

You might also tally the distributions of points won by players A and those won by players B for all the games. The most frequent distributions of points are likely to be close to 10 points for A and 5 points for B—that is, (10, 5), (10, 6), and (10, 4), since A scores approximately twice as often, although other combinations will certainly occur.

Conclude by asking the students to think of a way to change the rules so that the game is more fair. Since player A is about twice as likely as player B to score a point on each round, one simple solution is to award player B two points whenever a difference of 3, 4, or 5 occurs, which counteracts player A's advantage. Students can also change the rules by redistributing the differences for which points are awarded among players A and B. They might, for example, assign points on the basis of whether the difference is even or odd.

The next activity, Test Guessing, helps students connect the ideas of sample space, probability of events, random variable, and probability distribution. The first and second settings are binomial events—that is, events with each trial having only two outcomes. The idea of a binomial event is not made explicit for students, but this activity lays the groundwork for future study.

Test Guessing

Goals

- Determine the sample space for a random experiment
- Analyze the empirical probabilities for the experiment
- Define random variables for a sample space by identifying the possible values
- Find the probability associated with each possible value of the random variable

Materials and Equipment

- A copy of the blackline master "Test Guessing" for each pair or triple of students

p. 104

Activity

Introduce the activity by asking the students if they have ever guessed on a true-or-false test. Most students will probably admit to having guessed on some questions, although they may not admit to having guessed on all the questions on a test. Ask them how they guessed. Some students may say that they looked for a pattern, such as alternating true and false answers. Other students may say that they just tossed a coin. Still other students may say that they just answered whatever came into their heads.

Create a "test" of three true-or-false questions to which you are certain that the students will *not* know the answers. The questions might cover very advanced content (e.g., advanced mathematics), or they might be written in a language that the students do not know. Ask the students to answer the three questions and explain how they decided which answers to choose. Then compile the students' answers. It is likely that about half the students will choose each of the options for each question.

Divide the students into pairs or triples, and distribute an activity sheet for each group to complete.

Discussion

Begin the discussion by asking the students to give their answers to question 3. Be sure that they understand that there are eight equally likely outcomes (C means correct; W means wrong): CCC, CCW, CWC, CWW, WCC, WCW, WWC, WWW. These eight values form the sample space of this situation. Of these outcomes, only one (CCC) gives a score of 70 percent or greater, so the probability of passing is 1/8.

One way to think about computing the probabilities is to assume that the guesses are independent, so, for example, the probability of CCW is the product of the probabilities of C, C, and W for individual questions. Since the probability of C (or W) on each question is 1/2, the probability of CCW is 1/2 • 1/2 • 1/2 = 1/8.

You can extend the discussion by asking the students to determine the possible values for the random variable "number of correct

Chapter 2: Probability Distributions

answers." (0, 1, 2, and 3) Then ask what the associated probabilities are. (1/8, 3/8, 3/8, and 1/8, respectively) The associated probabilities constitute the probability distribution for this random variable. The only value that is associated with a passing rate of 70 percent is "three correct answers."

Once you are confident that most of the students understand the analysis for a three-question test, ask one or two students to explain how they answered questions 5 and 6. For question 5, the sample space has 32, or 2^5, elements: CCCCC, CCCCW, CCCWC, CCCWW, and so forth. The analysis proceeds essentially the same way, except that more elements must be considered. Achieving a score of 70 percent or greater occurs for six elements of the sample space: CCCCC, CCCCW, CCCWC, CCWCC, CWCCC, and WCCCC. The probability of passing, then, is 6/32.

For question 6, $P(C) = 1/3$ and $P(W) = 2/3$ for each question. The eight elements of the sample space are, therefore, not equally likely. For example, $P(CCC) = 1/3 \cdot 1/3 \cdot 1/3 = 1/27$, and $P(CWW) = 1/3 \cdot 2/3 \cdot 2/3 = 4/27$. The probability of passing with a score of 70 percent or greater is, then, 1/27.

This activity requires a fairly detailed examination of the sample space. The next activity, Number Golf, highlights the usefulness of the probability distribution in making decisions. In this case, the decisions involve choosing between pairs of game moves.

Number Golf

Goal

- Apply knowledge of a probability distribution to the development of a game strategy

Materials and Equipment

- A copy of the blackline masters "Number Golf: One Die" and "Number Golf: Two Dice" for each student
- Two dice for each group of students

pp. 105, 106

Activity

Divide the students into groups of two to four to play "number golf." Distribute the activity sheets, read the rules of the game, and allow the students to begin play. You may wish to make a few sample tosses of the dice to demonstrate the game.

The point of the game is to help students see that knowledge of a probability distribution can be useful for choosing which move to make in a game. When one die is used, the probability distribution is uniform, indicating that the probability of each outcome is the same. Being 2 away from the goal number is no advantage over being 6 away from the goal number. The simplest strategy to use in playing the one-die game, then, is to get as close to the goal number as possible. The game has been shown to be effective in helping students learn the underlying ideas (see Bright, Harvey, and Wheeler [1985]).

When two dice are used, the probability distribution of the sums is *not* uniform; the distribution is mound-shaped. The sums in the middle of the range are more likely to occur than the sums at the extremes. Since the probability of getting a sum of 7 is greater than that of getting a sum of 2 or 12, a player is more like to win if the cumulative sum differs from the goal number by 7 than if the cumulative sum differs from the goal number by 2 or 12. The best strategy to use in playing the two-dice game is to keep the cumulative sum as close as possible to either (goal number + 7) or (goal number – 7).

A *uniform distribution* appears flat (see fig. 2.8). Every outcome is equally likely to occur.

A *mound-shaped distribution* looks like a hill (see fig. 2.9). The outcomes in the middle of the range are the most likely to occur, and the outcomes at the extremes are the least likely to occur.

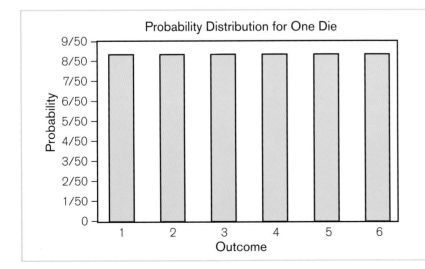

Fig. **2.8.**

A uniform distribution

Chapter 2: Probability Distributions

Fig. **2.9.**
A mound-shaped distribution

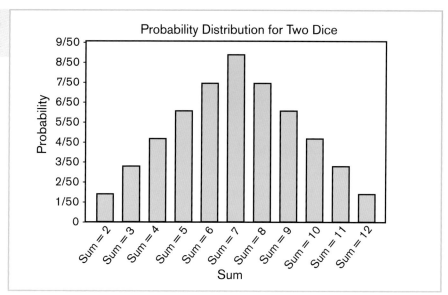

Discussion

The main goal of the follow-up to this game is to allow the students to explain their strategies for playing the game. Being explicit about their strategies will help the students understand how the mathematics that they know can be connected to activities in the real world.

For the one-die game, most students will probably be able to explain that they simply moved toward the goal number on each play. For the two-dice game, however, some students may not be able to explain their strategy explicitly. Be alert for any explanation that reflects an understanding of the mound-shaped distribution of the sums. You may need to paraphrase the students' explanations to make them clearer to the other students.

The next two activities, First Head and Strings of Heads, help students overcome some erroneous intuitions about the outcomes of coin tosses. In First Head, some students may be surprised that occasionally they must toss a coin numerous times before the first head appears.

First Head

Goals

- Calculate relative frequencies of events
- Compare relative frequencies with theoretical probabilities

Materials and Equipment

- A copy of the blackline master "First Head" for each student
- One coin for each student
- Paper and tape to create "biased" coins

p. 107

Activity

Introduce the activity by asking the students to predict how many times they would need to toss a coin until a head appeared. Most students will likely say one or two times. Then ask if anyone thinks that sometimes they might have to toss the coin four or five times before the first head appeared. Some students may say that could not happen. Do not try to reach a consensus at this point. Distribute the activity sheets to each student, and ask the students to carry out the activity individually. You may want them to compare their answers with those of a partner before you discuss the activity with the entire class.

Discussion

The probability of a head (H) on the first toss is 1/2, and for each subsequent toss, the probability decreases by half; that is, P(first H on second toss) = 1/4, P(first H on third toss) = 1/8, P(first H on fourth toss) = 1/16, and so on. Students are often surprised that the probability of no heads in ten tosses is $(1/2)^{10}$, which is the same as P(first H on tenth toss).

After the students have finished the activity sheets, ask a few of them to share their results for the fair coin. Then pool the data for the entire class, and compare the relative frequencies for the pooled data with those for the individual data. The relative frequencies of the pooled data should be fairly close to the theoretical probabilities, whereas the students' individual data may show quite a range of values. Depending on the level of understanding of your students, you may choose to discuss how the theoretical probabilities are computed or to limit the discussion to the data alone.

The data for the biased coins will vary, since the results depend on how much paper students tape to the coins and the kind of coins used. That is, the same amount of paper taped to a penny and a quarter will cause different degrees of bias. Since the conditions of the students' experiments for the biased coin are not identical, it is not reasonable to pool the data. The probability distributions for the coins with different degrees of bias are different, and data from different distributions should not be pooled. Ask several students to report their results for the biased coin and compare their individual results for the fair and biased coins.

The activity Strings of Heads further challenges students' biases about the outcomes of coin tosses. Many students may expect that the maximum length of runs of heads or tails will be small, say, two, three, or four tosses. Having runs as long as five tosses will seem "wrong" to those students.

Strings of Heads

Goals

- Calculate the relative frequencies of events
- Interpret relative frequencies as probabilities

Materials and Equipment

- A copy of the blackline master "Strings of Heads" for each group of students
- A fair coin for each student
- Win-and-loss records of baseball or basketball teams (supplied by students)
- An overhead projector

p. 109

Activity

Introduce the activity by asking the students what they think a *run of heads* or a *string of heads* in a series of coin tosses means. If the students cannot give accurate definitions for the phrases, explain that they mean two or more consecutive heads in a series of coin tosses. Put an example (e.g., HHTHHHTTTT) on the overhead projector, and ask the students what the runs of heads are in the example. There is one run of two heads and one run of three heads.

Then ask what the longest string of heads is likely to be in a series of ten tosses. Most students will probably suggest that two, three, or perhaps four heads would be the longest. If you ask the students to explain their answers, they may say something like "Since heads and tails are equally likely, heads and tails usually alternate, so you won't get longer strings of heads." Do not try to correct this misconception at this point in the activity.

The students should share the work in pairs or small groups. You may want the students to compare their answers with those of another group before you discuss the activity with the class as a whole. Allow forty to sixty minutes for this activity.

Discussion

For the ten-toss situation, ask two or three groups to share their results. For any group, it is quite likely that the estimates of some probabilities (i.e., the relative frequencies) will be 0 if, for example, the group did not observe any set of tosses in which the longest string of heads was three. Ask whether these zero values are likely to represent the true probabilities. Then discuss the differences in the estimates of probabilities among the groups of students that share their results. The estimates are likely to show considerable variability. Finally, pool the data for the class as a whole and compute the estimates of probabilities. The pooled estimates can be expected to be much closer to the theoretical probabilities than the estimates of the individual groups are. The data in table 2.4 were obtained from a simulation of one hundred repetitions of ten tosses of a fair coin.

Table 2.4
Frequencies of Different Strings of Heads in One Hundred Repetitions of Ten Tosses of a Fair Coin

Length of Longest String	Frequency
1	10
2	33
3	33
4	13
5	7
6	1
7	1
8	2
9	0

Repeat the discussion for the twenty-toss situation. Discuss the changes in probabilities for the common events (e.g., length of longest string = 1 or 2 or 3 or ... or 10). Since there are more events in this situation, the probabilities will obviously change, but the probabilities of the new events (e.g., length of longest string = 15) are all very small, so the changes are perhaps not as great as the students might expect. The data in table 2.5 were obtained from a simulation of one hundred repetitions of twenty tosses of a fair coin.

Table 2.5
Frequencies of Different Strings of Heads in One Hundred Repetitions of Twenty Tosses of a Fair Coin

Length of Longest String	Frequency
1	2
2	13
3	26
4	27
5	20
6	3
7	4
8	4
9	1

The two sets of simulated data display some clear differences in the frequencies for the different lengths of strings. It is interesting that there were no runs greater than 9 in the second simulation.

Finally, conduct a similar discussion of the sports situation. Since the probability of W is not likely to be 1/2 for each game, the distribution will look more like that of a "biased coin" than a fair coin.

Conclusion

The activities in this chapter help students explore both theoretical and empirical probabilities. The next chapter makes explicit the law of large numbers: As the number of observations increases, the empirical probabilities should approximate the theoretical probabilities.

Kader and Perry (1998) describe a game that allows a further exploration of runs. Players score "hits" and "misses" as they push pennies across a game board. In this game, the runs are influenced by the skill of the players.

NAVIGATIONS SERIES

GRADES 6–8

NAVIGATING through PROBABILITY

Chapter 3
Prediction and the Law of Large Numbers

Important Mathematical Ideas

One of the faulty intuitions that students may bring to the study of probability is that if an expected outcome has not occurred for several trials of an experiment, then that outcome is more likely to occur on the next trial. For example, if a coin is tossed four times and tails turns up each time, many people will say that heads is more likely to occur than tails on the fifth toss. People will often refer to this prediction as "the law of averages," but mathematicians call it "the gambler's fallacy." Obviously, the coin does not remember what happened on the first four tosses, and it does not redistribute its mass so that heads is more likely to occur on the fifth toss. Rather, each toss is independent of every other toss, and the probability of heads or tails does not change in response to previous outcomes. Nonetheless, this intuition is very persistent and influences many people's ideas of probability.

Most people would like to be able to predict such particular events as what will happen on the next toss of a coin. Being able to make such predictions would give people considerable control over their world. However, the ideas of probability deal more with predictions over the long term than with predictions of individual outcomes.

Part of the difficulty in internalizing ideas of probability is that people are inclined to look for order in the world, especially in the world of mathematics. The notion that random events are a fundamental part of some mathematical situations is often difficult for students to understand. For example, recognizing and characterizing patterns is an

"The idea that individual events are not predictable ... but that a pattern of outcomes can be predicted is an important concept that serves as a foundation for the study of inferential statistics."

(NCTM 2000, p. 51)

important part of mathematics. In algebra, patterns are often represented by formulas, which show relationships among variables. The search for patterns, however, may not be appropriate when data represent random events. Indeed, trying to impose patterns on random events can lead to incorrect reasoning. Students need to know when patterns can be expected and when patterns should not be expected.

Probability, then, is in many ways unlike other branches of mathematics. In particular, students have to accept that the long-term trends in events are more important than individual outcomes of events. Students have to give up the expectation of finding quick, definitive answers and accept that in the study of probability, it is important to look at the big picture, accept some uncertainty in answers, and try to avoid imposing regularity on data that may not be regular.

Populations and Samples

One reason that probability is difficult for some students is that many probability contexts involve imagining infinitely many outcomes. For example, even in the apparently simple situation of tossing one coin, with each toss yielding one of two outcomes—heads or tails—the coin could theoretically be tossed forever, and it is obviously impossible to list all the outcomes of all those tosses. It may be difficult for some students to reconcile their understanding of the infiniteness of all outcomes of all possible tosses with the very finite set of outcomes for each individual toss.

When students conduct an experiment, they are examining only a small subset of the set of all possible outcomes that might occur if the experiment were repeated. The set of all possible outcomes is called the *population*, and the smaller set of outcomes generated by the experiment is called a *sample*. It is important that students understand the difference between a population and a sample drawn from that population. When the outcomes of an experiment are used to make predictions about the population, the challenge is to ensure that the sample is representative of the entire set of all possible outcomes.

The concepts of sample and population are part of both probability and data analysis, although students may not immediately see the connection. For example, electronics manufacturers may want to know what percent of households in the United States have a DVD player. It is clearly impractical to survey every household, but by surveying a carefully constructed sample of households, manufacturers can make a very accurate estimate of the percent of all households that have a DVD player. The process of creating a representative sample for the U.S. population is complex, but middle-grades students can begin to consider the underlying ideas. The main point is that in general, the larger the sample, the greater the confidence in the inferences made from the data in that sample.

An experiment will generate a useful sample only when trials are performed identically and independently—that is, when repetitions of the experiment occur under the same, or nearly the same, conditions. Sometimes the differences in the distributions of data generated by students' experiments can be attributed to differences in the ways that they performed the experiments. If students are drawing chips out of a bag, the chips that are at the top of the bag may be more likely to be drawn.

> An *experiment* is a process for gathering data (e.g., repeatedly tossing a coin). In probability settings, an experiment must be repeatable under essentially the same conditions. A *population* is the complete set of objects or events about which data can be gathered (e.g., all the possible tosses of a coin). A *sample* is any subset of a population (e.g., one hundred tosses of a coin).

If, then, the chips are not mixed thoroughly before each draw, the sample may not be random.

One way to get a "fair" sample—or at least to avoid the appearance of an inappropriate sample—is to use a random sample. In standard probability experiments, a random sample can be obtained by mixing chips thoroughly, by using spinners that spin freely, or by rolling dice that are not "loaded." As long as the technique is objective or fair, the resulting sample will probably be considered fair, or random.

In the real world, choosing representative samples can be complicated. For example, *Entertainment Weekly* asks users of its Web site to rate movies. The people who submit ratings must of necessity have access to the Internet and may be particularly interested in expressing their opinions. Therefore, they may or may not be representative of all moviegoers; it is impossible to know. But it is important to recognize that there is at least a possible bias in the sample that might make the ratings unrepresentative of the view of all moviegoers.

A *simple random sample* is a sample that is chosen in such a way that any subset of that size has the same probability of being chosen as any other subset of the same size; for example, a sample composed of three chips from a bowl of thoroughly mixed chips has the same probability of being chosen as any other sample of three chips. In a probability experiment, a simple random sample gives us no reason to believe that bias exists in the way that outcomes are generated.

The Law of Large Numbers

The law of large numbers is a fundamental principle in statistics and probability. Stated simply, it says that for a given event, the experimental, or empirical, probability (the probability generated from experimental data) is more likely to approximate the theoretical probability (the probability generated from a mathematical analysis of the context) as the sample size increases. In other words, large samples are more likely to reflect, or be representative of, the parent population than small samples. Small samples may produce experimental probabilities that differ markedly from the parent distribution. For example, it is not unusual for two tosses of a coin to result in two tails, but that result does not mean that we should conclude that tossing a coin always produces tails.

People often assume that the theoretical probabilities will be reflected in every sample, no matter how large or how small. In a classic article, Kahneman and Tversky (1972) used the phrase *law of small numbers* to refer to the incorrect assumption that every sample, no matter its size, reflects the characteristics of the parent population. The law of large numbers holds that this assumption is false; it is only in the long run that we can expect empirical probabilities to approach the theoretical probabilities. Unfortunately, however, the law does not state how large a sample is needed.

An example of sampling is the weekly rating of television programming conducted by Nielsen Media Research, which reports the most-watched television programs in the United States. The results are based on a carefully selected sample of households whose viewing habits are monitored. From this sample, inferences are drawn about the viewing habits of the general population of TV viewers. As the size of the sample increases, so does the likelihood of making accurate inferences from the sample. The concept of sampling is central to data analysis, but some students may not understand that more-accurate inferences can be drawn from large sets of data or that small samples are likely to produce results that seem unusual.

One of the common misconceptions of the law of large numbers is that if one outcome does not occur for several trials, then it is more

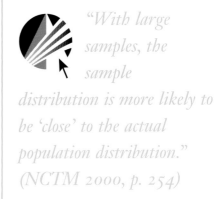

"With large samples, the sample distribution is more likely to be 'close' to the actual population distribution." (NCTM 2000, p. 254)

likely to occur on the next trial. The law of large numbers states that over the long term, or on average, the characteristics of a large sample will begin to approach the characteristics of the population from which the sample is taken. It does not say anything about any particular trial or any particular element of that sample.

For instance, suppose that a two-color chip—let us consider, as an example, a chip that is red on one side and yellow on the other—is to be tossed ten times. Before any tosses are made, most people would predict that about five of the tosses would be red and about five would be yellow. Suppose, however, that the first four tosses are red. Should we stick to the original predictions for the set of ten tosses? Probably not. For the remaining six tosses, we would expect about equal numbers of red and yellow outcomes, or about three of each color. At the end of this particular set of ten tosses, therefore, given that the first four are red, we might expect seven red outcomes. The outcomes of the last six tosses are not influenced by the outcomes of the first four tosses, even though those outcomes seem somewhat unusual.

The law of large numbers also provides a rationale for pooling data: The limited data gathered by an individual student, for example, may not reflect the population very well because the data set is too small. By pooling data, we create a larger sample, and we therefore have greater confidence in the inferences drawn from it. Of course, the data must be gathered under essentially identical circumstances. For example, one student might roll a fair die six times and get three 2s (50% of the rolls) and no 5s (0% of the rolls). Another student may get three 5s (50% of the rolls) and no 2s (0% of the rolls). Neither of these small samples produces surprising results, but neither one is a good representation of the population of all possible rolls of a die. Pooling data from twenty-five sets of six rolls produces a sample of 150 rolls, in which it is unlikely—though not impossible—that any single value would occur 50 percent of the time. The characteristics of the population—in this example, that each event is equally likely—would be more evident in the large sample than in either of the small samples.

Large and Small Samples

Sometimes students overapply notions of proportionality in dealing with large or small samples. For example, consider whether it is more likely for 75 percent of coin tosses to be heads in a sample of four tosses or in a sample of forty tosses. Some students will argue that the likelihood of 75 percent of outcomes being heads is independent of the sample size. Their reasoning may be an artifact of the emphasis on proportional reasoning common in middle-grades curricula, but in the context of sampling in probability, this reasoning is incorrect. Small samples are much more likely than large samples to yield unusual results, and this concept can be understood by appealing to the law of large numbers. (This idea is explored in the activity Two Hospitals.) If the coin is fair, the population of all tosses will result in 50 percent heads and 50 percent tails, so the result of 75 percent heads does not align well with the characteristics of the population. The law of large numbers suggests that the characteristics of the larger sample are more likely to align with

"Students could be asked to predict the probability of various outcomes ... [and] discuss whether the results of the experiment are consistent with their predictions."

(NCTM 2000, p. 254)

the population characteristics. In this instance, the smaller sample's yielding three heads out of four tosses is less unexpected than obtaining thirty heads out of forty tosses for the larger sample.

Another example of this idea is illustrated by students' guessing on true-or-false tests. Are they more likely to make a score of 100 percent by guessing when a test consists of four questions or forty questions? Most people would conclude—correctly—that a perfect score is more likely in the four-question situation. That is, it is more likely to get an unusual result for a smaller sample than for a larger sample.

One of the reasons that exploring large and small samples is so important is that in some situations, the probability of an event can be determined only through data collection. For example, the likelihood of a randomly selected person's having the blood type A-positive is determined only by collecting data from the general population. In order to be confident that we know the proportion of all people with A-positive blood, it is important to recognize that larger samples are more likely than smaller samples to accurately reflect the characteristics of the population.

In all instances of prediction, however, it is important to understand that an inaccurate prediction is not the same as an "erroneous" prediction. Predictions are not fact, but when they are grounded in quantitative thinking, they can be truly educated guesses. The process of formulating a prediction in which we have confidence is more important than the accuracy of the prediction itself. The legitimacy of the prediction process should not be judged by the accuracy of the prediction. The ways in which students explain the outcomes of experiments and make predictions about future events will reveal much about how they are thinking about probability. It is important that teachers help students focus on global patterns of outcomes rather than on the outcomes of individual trials.

What Might Students Already Know about These Ideas?

When students make judgments about probability, their experiences in rolling a pair of dice can be either beneficial or detrimental. For example, playing a board game may help students learn that some outcomes, such as a sum of 7, are more likely to occur than others, such as sums of 2 or 12. Conversely, a student may recently have won a game by rolling "doubles" and then assert that he or she is "good at rolling doubles." When people make predictions about the likelihood of an event on the basis of recent experiences, they tend to rely on *availability*, or *recency* (Shaughnessy 1992). This powerful heuristic can cause faulty reasoning and erroneous predictions.

The activity Racing Game can help teachers determine what their students understand about how the distribution of the sums of the values on dice changes as the number of tosses of the dice increases. The probability of winning is better approximated as the size of the set of trials (i.e., the size of the sample) increases. This activity also reveals information about how students formulate predictions.

Racing Game

Goals

To assess students'—

- ability to determine the fairness of a game;
- understanding that better inferences can be drawn from larger samples than from smaller ones.

Materials and Equipment

- A copy of the blackline master "Racing Game" for each student
- A pair of dice for each group of students
- Transparency copies of the blackline master "Pooled Racing-Game Data"

Activity

Before beginning to play the game, ask the students, "How do you know if a game is fair?" Have them justify their thinking. Some students may say that a game is fair if each player wins equally often, and other students may argue that a game is fair if no player has an advantage. The latter view is equivalent to saying that every player has the same probability of winning, although you should not expect all the students to use such terminology. A few students may believe that a game is unfair when they do not win. This rationale shows a very egocentric kind of reasoning that has no basis in mathematical ideas.

Divide the students into groups, and give each student an activity sheet. Have the students read the instructions for the game on the activity sheet, and ask if they think the game is fair. Do not at this point ask for rationales for their answers. Some students may erroneously believe that each car has the same chance of winning and indicate that they think this game is fair.

Having students work in small groups will increase the pace of the game. You may want the students to compare their answers with those of another member of the group before you discuss their work with the whole class.

Discussion

Ask several students to present and explain their initial predictions (questions 1 and 2). For example, some students who predict that car 7 will be the winner may be doing so because 7 is perceived to be a "lucky number" or because "7 is in the middle" rather than because they understand that the sum 7 has a greater probability of being rolled than any other sum. It is important that the students hear a variety of rationales and that none of their suggestions be ridiculed by the other students.

Then discuss which car was leading at each of the mileposts. The leaders at each milepost during fourteen computer simulations of the game are given in table 3.1. The leaders for the early mileposts are likely to be more diverse than the leaders for the later mileposts. This likely result reflects the law of large numbers: The results of small

pp. 113, 115

Samples and descriptions of the full range of students' probabilistic reasoning can be found in Jones, Thornton, Langrall, and Tarr (1999).

"To correct misconceptions, it is useful for students to make predictions and then compare the predictions with actual outcomes."
(NCTM 2000, p. 254)

samples (e.g., the leader after one or two rolls) may not reflect the characteristics of the population so well as the results of large samples (e.g., the leader after eight or nine rolls). Ask the students to explain why the list becomes less diverse for the later mileposts. Their explanations, or their inability to suggest explanations, will reveal much of their thinking about small and large samples. For example, some students may explain the diversity of outcomes at milepost 1 by saying, "With one roll, anything can happen!" Others may correctly reason that the list becomes less diverse because "the race cars with the best chance of winning will eventually pull away from the others."

Table 3.1
Race Cars Leading at Each Milepost in Fourteen Simulations

Game Number	Leader at Milepost									
	1	2	3	4	5	6	7	8	9	10
1	8	7	7	5	6	7	5	7	7	7
2	2	6	9	5	7	7	7	8	7	8
3	4	9	5	7	6	6	6	7	8	7
4	10	3	6	9	8	9	8	6	9	6
5	3	5	7	6	5	9	7	9	6	7
6	6	7	7	8	5	8	7	5	6	7
7	11	10	10	7	7	5	9	7	8	7
8	12	8	8	7	7	5	9	8	7	7
9	9	3	8	8	9	7	6	6	7	8
10	8	7	7	6	9	6	7	7	8	7
11	7	6	6	10	7	8	7	8	9	6
12	5	10	7	6	8	7	8	8	5	7
13	10	6	5	8	6	7	7	6	7	7
14	7	4	9	7	9	6	6	7	6	8

An interesting discussion can be prompted by the question "After which milepost did you begin to see trends in the data?" Some students may notice cars 6, 7, and 8 "pulling away" after milepost 3 or 4, whereas other students may not begin to see the trends until milepost 7 or 8. The data from the computer simulation (see table 3.2) show a narrowing of the range of the leading cars at each milepost. The range narrows at mileposts 2, 3, 5, and 10.

Pool the class's data by transferring the data from the score chart of one student in each group onto transparency copies of the "Pooled Racing-Game Data" form. Each transparency will hold the data from

Table 3.2
Range of Cars Leading at Each Milepost

Milepost	Range of Leading Cars
1	2–12
2	3–10
3	5–10
4	5–10
5	5–9
6	5–9
7	5–9
8	5–9
9	5–9
10	6–8

The probability that car 6, 7, or 8 will win the "racing game" is approximately 81 percent. This probability is based on a racetrack of length 10 and was determined from 1,000,000 trials of a computer simulation of Racing Game with Two Dice, one of the applets on the accompanying CD-ROM.

four games. Lay the transparencies on the overhead projector one after another, and point out that the charts are like line plots. Ask the students to look for similarities in the data displays. Ask the students to explain how the graphs are similar and why they all have roughly the same shape. Focusing on the shapes of the displays helps the students begin to see trends in the data. When data from many games are combined, the sample size is increased, and the distribution of the outcomes begins to appear stair-stepped and symmetric, like that in figure 3.1, which displays data generated from a computer simulation.

Fig. **3.1.**

Racing-game-simulation data

Car 2	Car 3	Car 4	Car 5	Car 6	Car 7	Car 8	Car 9	Car 10	Car 11	Car 12
					X					
					X					
					X					
					X					
					X					
					X					
					X	X				
					X	X				
				X	X	X				
				X	X	X				
				X	X	X				
				X	X	X				
				X	X	X				
				X	X	X				
				X	X	X				
			X	X	X	X				
			X	X	X	X				
			X	X	X	X	X			
			X	X	X	X	X			
			X	X	X	X	X			
			X	X	X	X	X			
			X	X	X	X	X			
			X	X	X	X	X	X		
			X	X	X	X	X	X		
			X	X	X	X	X	X		
			X	X	X	X	X	X		
		X	X	X	X	X	X	X		
		X	X	X	X	X	X	X		
		X	X	X	X	X	X	X		
		X	X	X	X	X	X	X		
		X	X	X	X	X	X	X	X	
		X	X	X	X	X	X	X	X	
		X	X	X	X	X	X	X	X	
	X	X	X	X	X	X	X	X	X	
	X	X	X	X	X	X	X	X	X	
	X	X	X	X	X	X	X	X	X	
	X	X	X	X	X	X	X	X	X	
	X	X	X	X	X	X	X	X	X	
	X	X	X	X	X	X	X	X	X	
	X	X	X	X	X	X	X	X	X	X
	X	X	X	X	X	X	X	X	X	X
	X	X	X	X	X	X	X	X	X	X
X	X	X	X	X	X	X	X	X	X	X
X	X	X	X	X	X	X	X	X	X	X
X	X	X	X	X	X	X	X	X	X	X
X	X	X	X	X	X	X	X	X	X	X
X	X	X	X	X	X	X	X	X	X	X
X	X	X	X	X	X	X	X	X	X	X

Since this activity is a preassessment, it is unreasonable to expect the students to understand the law of large numbers. However, their explanations may incorporate parts of this principle. For example, some students may say that since each transparency shows four games, the shapes should be the same. This explanation implicitly acknowledges that four games include more data than one game, but you may have to probe students' thinking to make this idea more explicit. Other students may say that since the sums 6, 7, and 8 are more likely to occur than the others, cars 6, 7, and 8 are more likely to move toward the finish line sooner. This explanation can be expanded by asking why the mound seems relatively higher for the display of four games than for the display of any individual game.

Similarly, a discussion of which cars are the farthest behind at each milepost can help the students see how increasing the sample size will tend to yield results that are more similar to the characteristics of the population. As the length of the race increases, the numbers of the cars that are the farthest behind tend to the extreme sums, namely, 2, 3, 11, and 12. This result is expected, since these sums are less likely to occur than the other possible sums.

To highlight how the length of the racetrack influences the probability of winning, use the applet Racing Game with Two Dice on the CD-ROM. Assign 2–12 to players A–K, respectively, and play according to the rule "Only lucky player moves." Enter "10" when asked "How long do you want the race to be?" and then click "Start the race." Next enter "1000" and click "Run automatically" to simulate 1000 races. Record the number of times racers 2, 3, 11, and 12 win the game. Change the length of the track to 4 units and repeat the simulation of 1000 races. In one experiment of this kind, racers 2, 3, 11, and 12 combined won 28 times out of 1000 runs on a racetrack of length 4, whereas the same racers won only 3 times out of 1000 on a racetrack of length 10. Can the students explain why?

You may want to have the students determine the theoretical probability of each car's advancing on a single roll. Representing the number pairs in a line plot may help the students draw connections between the theoretical and experimental probabilities, since the displays in figures 3.1 and 3.2 (a display of the shape of the possible number combinations for a roll of two dice) should have a shape very similar to the shape of the experimental data on the "Pooled Racing-Game Data" transparency.

Other games are presented by Bright, Harvey, and Wheeler (1981) and Van Zoest and Walker (1997). Both articles are available on the CD-ROM.

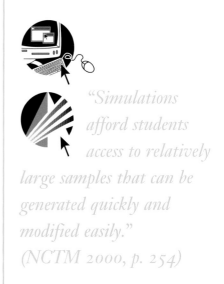

"Simulations afford students access to relatively large samples that can be generated quickly and modified easily." (NCTM 2000, p. 254)

Fig. **3.2.**

The possible number combinations for rolls of two dice in "Racing Game"

					(6, 1)					
				(5, 1)	(5, 2)	(6, 2)				
			(4, 1)	(4, 2)	(4, 3)	(5, 3)	(6, 3)			
		(3, 1)	(3, 2)	(3, 3)	(3, 4)	(4, 4)	(5, 4)	(6, 4)		
	(2, 1)	(2, 2)	(2, 3)	(2, 4)	(2, 5)	(3, 5)	(4, 5)	(5, 5)	(6, 5)	
(1, 1)	(1, 2)	(1, 3)	(1, 4)	(1, 5)	(1, 6)	(2, 6)	(3, 6)	(4, 6)	(5, 6)	(6, 6)
Car 2	Car 3	Car 4	Car 5	Car 6	Car 7	Car 8	Car 9	Car 10	Car 11	Car 12

Selected Instructional Activities

The remaining activities in this chapter lead students to explore further the law of large numbers and the making of predictions from data sets of different sizes. There is no particular order to the activities, although The Long Flight Home is likely to be the most challenging situation for students to analyze accurately.

The next activity, Two Hospitals, gives students an opportunity to explore intuitions about how sample size and results are related.

Two Hospitals

Goals

- Understand that small samples may yield unusual or unexpected results
- Understand that distributions representing larger samples are more likely to reflect the parent distribution than distributions that represent smaller samples
- Recognize that the experimental probability approaches the theoretical probability of an event as the size of the sample increases

Materials and Equipment

- A copy of the blackline master "Two Hospitals" for each student
- One two-color chip (or a coin) for each student
- An overhead projector and transparencies
- Half-centimeter grid paper (available on the CD-ROM)
- Calculators (optional but helpful)

p. 116

Activity

Distribute grid paper and a copy of "Two Hospitals" to each student, and review with the students the statement of the problem. Be sure that they understand that one difference that could be expected between the birth data generated by the large hospital and those generated by the small hospital is the sizes of the data sets. Some students may explain that both samples represent at least 80 percent female births. You may want to survey the students about question 1 before they begin their work. Although the proportions of female births mentioned in the question are the same, many students may erroneously assume that the two events have the same probability. Others may reason that at least twenty female births out of twenty-five births is the more probable result, since six events comprise it—namely, 20, 21, 22, 23, 24, and 25 female births. Only two events satisfy the conditions for the smaller set—namely, 4 and 5 female births. Acknowledge all the students' explanations without indicating which, if any, are correct.

You may need to explain the third column of the table, "Experimental Probability." As an example, suppose that the first five outcomes are M, F, F, M, and M. Then the proportion of females is 0/1, 1/2, 2/3, 2/4, and 2/5, respectively, yielding *experimental probabilities* of .00, .50, .67, .50, and .40, respectively. Calculators, if available, will help the students in this task.

The students should complete the activity sheets individually. You may want them to compare their answers with those of a partner before you discuss their work with the whole class.

Discussion

Students often share the common misconception that samples of any size should be representative of the population from which they are drawn. For repeated tosses of a coin, for example, there is a pervasive

Chapter 3: Prediction and the Law of Large Numbers

"Misconceptions about probability have been held not only by many students but also by many adults."
(NCTM 2000, p. 254)

belief that every sample should contain approximately the same number of heads and tails. Some students believe erroneously that an unusual number of, say, heads is equally likely to occur in samples of any size. Of course, that is not so. Be alert to this misconception as students discuss their work.

The amount of follow-up required for questions 3, 4, 6, and 7 will depend in part on what you observe as the students work. If the students have a great deal of trouble answering these questions, then class time may be wisely used in discussing how they generated their answers.

It is important, however, to discuss the data generated for questions 5 and 8. For question 5, you can create a tally on the board or the overhead projector of the number of students who found at least four females (i.e., four or five females) among the simulated five births. For question 8, you can create a tally of the number of students who found at least twenty females (i.e., twenty, twenty-one, twenty-two, twenty-three, twenty-four, or twenty-five females) among the simulated twenty-five births. You would expect more students to have found four or more females out of five births than found twenty or more females out of twenty-five births, although it is always possible that the results will not be as you expect. If the results are unexpected, you may need to question how the students gathered their data.

The probability of at least four female births out of five births is .1875—that is, $P(4 \text{ female births}) + P(5 \text{ female births}) = 5 \cdot (1/2)^5 + (1/2)^5$, so in a class of twenty-five students, you can expect about four to six students to have simulated at least four female births in five births. The probability of at least twenty females out of twenty-five births is about .0021 (determined by evaluating the binomial expansion), so it is *very* unlikely that *any* student will have simulated this event. The results of the experiments should stimulate students' interest, so you can ask, "Why do you think no one obtained twenty or more females births? Would you expect different results if we carried out the simulation again?" Some students may believe that it was merely luck that no one obtained the desired result, and they may assert that a second simulation of twenty-five births would be likely to yield results similar to the results of the simulation of five births. Others may argue that their data "prove" that obtaining at least 80 percent female births is more likely in the smaller sample. Neither argument is complete, and the mathematical knowledge needed to make a complete argument is beyond the middle-grades curriculum.

Next, discuss the graphs that the students made. A graph of twenty-five values generated by a simulation is given in figure 3.3. The students' examples will certainly look different, although they would be expected to share the characteristic that the graph becomes more level as the number of trials increases.

You can initiate a discussion of the graphs by asking the students to describe the general shape of their graphs or by asking why there is so much vacillation in the first few trials. The students should have noticed that the "up-and-down pattern" in the graph tends to even out as the number of trials increases. This observation is central to understanding the relationship between experimental and theoretical probability as supported by the law of large numbers.

Bright and Hoeffner (1993) present another example of this phenomenon.

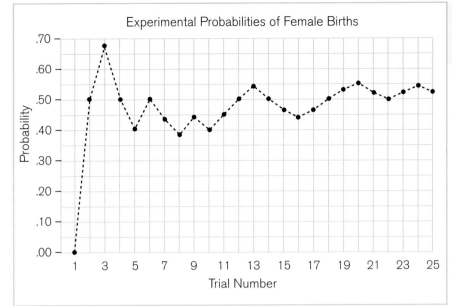

Fig. **3.3.**

A sample coordinate graph for the "Two Hospitals" data

If time allows, pool several sets of data, and ask the students to make similar graphs for one hundred or more trials. The right part of the graphs will almost certainly "settle down" close to the theoretical value. If some students continue to struggle with the notion that data from large samples are more likely to be representative of the data in the parent distribution, you could ask, "Are you more likely to observe 100 percent female babies for 2 births, 20 births, or 200 births?" It is clear that a larger sample is less likely to yield an unusual ratio of female births; this realization may make the law of large numbers more apparent to the students.

The next activity, Gumball Machine, also focuses students' attention on the differences between small samples and large samples. For some students, the context of Gumball Machine may be more accessible than that of Two Hospitals, and the simulation and the use of chips to represent the gumballs should be familiar to the students.

Chapter 3: Prediction and the Law of Large Numbers

Gumball Machine

Goals

- Understand the concept of independent events
- Understand that events associated with small probabilities can and do occur
- Realize that the distribution of data from small samples often does not reflect the parent distribution

Materials and Equipment

p. 119

- A copy of the blackline master "Gumball Machine" for each student
- A paper cup for each pair of students
- Ten colored chips (four blue, three yellow, two green, and one red) for each pair of students

Activity

Distribute a copy of the activity sheet to each student. Introduce the activity by having the students read question 1; then ask, "If I were to close my eyes and select one gumball at random, which color do you predict would be drawn?" Some students will predict "blue because it has the most," and "blue" is the best prediction of a single outcome. Others may use more-primitive reasoning; for example, some may predict "red," since it is their favorite color.

Then ask which of the following statements is true:

1. Blue is likely to be drawn.
2. Blue is likely not to be drawn.
3. Blue is equally likely to be drawn or not to be drawn.

A more detailed discussion of students' reasoning can be found in Jones, Thornton, Langrall, and Tarr (1999), which is available on the CD-ROM.

This question gets at the important notion of *complementary events*. Blue either *will* or *will not* be selected. In this example, it is more likely that blue will not be drawn, yet many students may argue that it is more likely that blue will be drawn. Their reasoning may reflect what is called the *outcome approach*. The students may be making pairwise comparisons of the likelihood of drawing blue to the likelihood of drawing each of the other colors. Since there are more blue gumballs, students often erroneously conclude that "drawing blue" is more likely than "not drawing blue." In doing so, these students neglect to see the event "not drawing blue" as the union of several disjoint events: drawing yellow, drawing green, and drawing red. It may be helpful for these students to describe the event "not drawing blue" as the single event "drawing any chip that is not blue." This representation conveys the fact that the odds are 6 to 4 against drawing blue.

Next, using one of the paper cups to represent the gumball machine and the chips to represent the gumballs, select one chip from the cup at random. Although the probability of selecting a blue chip is 40 percent, it is likely that a different color will be drawn. Drawing a yellow, green, or red chip will demonstrate to the students a fundamental idea in probability—namely, that events associated with relatively small probabilities *can* occur. Moreover, the selection of one of these colors supports the idea that the event "not drawing blue" is more likely than its complement. You may want to repeat this demonstration several times to reinforce these important ideas.

The students should work either individually or in pairs. You may want them to compare their answers with those of a partner before you discuss the activity with the entire class.

Discussion

Students hold pervasive beliefs that often conflict with the concept of independence. For example, although teachers understand that sampling with replacement represents independent trials, students may impose regularity on a random phenomenon by adopting any one of several heuristics. As an example, suppose that a yellow gumball is drawn twice in a row. In this situation, students might use *positive recency* to predict that the probability of selecting a yellow gumball has increased, since "yellow keeps getting drawn." These students are likely to predict that a yellow gumball will be drawn next on the basis of only a small number of trials. In contrast, after observing the selection of two consecutive yellow gumballs, other students may use *negative recency* to predict that yellow will not be selected on the next draw. These students believe that the probability of drawing a yellow gumball has decreased in a with-replacement situation when, in reality, the probabilities of all events remain unchanged, since the sample space is preserved. That is, because the gumball is replaced after each draw, the sample space has been restored to its original state.

Although in any given selection, yellow is indeed more likely not to be drawn than to be drawn, some students may use reasoning that is not mathematically grounded. For example, after two yellow gumballs have been drawn, some students might argue, "Yellow probably won't get picked, since it is really hard to draw yellow three times in a row." A stronger argument is that the odds are 7 to 3 against the selection of yellow in any given draw. Still other students may erroneously believe that every event and its complement are equally likely and reason accordingly that there is a "50-50 chance" of selecting yellow in any given draw. Asking the students to decide which of these—or other—explanations is correct can provoke a rich whole-class discussion.

After the students have finished the activity sheets, create a tally to determine which color occurred most often in the initial ten draws. Given the small sample size, it would not be unusual for some students to have selected yellow—or possibly green—most often, although it would be unlikely for red to have been drawn most frequently. Smaller samples are more likely to yield unusual results, and if even one student's simulation produced a color other than blue most often, you could use that result as the focus of a whole-class discussion. You might initiate the discussion by saying something like, "Most of you predicted that blue would be drawn most often, and yet yellow was selected most frequently by Maria." Then you could ask, "How can you explain this outcome?" Larger samples are likely to reveal which color is likely to occur most often; in fact, the relative frequency of each color drawn becomes more predictable as the sample size increases.

Positive recency refers to the belief that an event is more likely to occur again because it has already happened. *Negative recency* refers to the belief that an event that has already happened is less likely to occur again.

Chapter 3: Prediction and the Law of Large Numbers

For those who struggle to accept the fact that any random sample of ten draws will not likely reflect a 4-3-2-1 distribution, pose the following scenario: "Suppose I begin sampling with replacement and my first nine random draws are four blue, three yellow, and two green gumballs. On the tenth (and final) draw, what color do you predict will be selected?"

Even though the sample distribution is on target to reflect the parent distribution precisely, the odds are 9 to 1 against selecting red on the tenth draw.

After the students have investigated which color of gumball was drawn most often, survey the class about the distribution of the outcomes of their simulations. Use students' intuition that a 4-3-2-1 distribution would occur to focus on the differences between their predictions and the results of their simulations. Ask, "Did anyone get four blue, three yellow, two green, and one red gumballs?" Take advantage of the fact that most students' samples do not reflect this distribution to ask them to explain the discrepancy. The students may offer "randomness" as a plausible explanation; that is, they may say that since the draws were random, the results were also random.

Repeat the follow-up discussion for the combined samples of twenty draws. In general, you can expect more students to have identified blue as having occurred most often for twenty draws than for ten draws, since a sample of twenty is likely to be more representative of the parent population.

If time allows, pool the data from the entire class. Have the students compute the experimental probabilities for each pooled event. In a sample this large, the proportion of blue, yellow, green, and red outcomes will likely approximate the .40-.30-.20-.10 distribution. Moreover, the pooled data should indicate the relationship between what *should* occur (theoretical probability) and what *does* occur (experimental probability) as the sample size increases.

The previous activities in this chapter have provided examples of how the characteristics of the population are better represented in a large sample than in a small sample. In some instances, however, we do not know what the characteristics of the population are. Students rely on the validity of the law of large numbers to generate approximate characteristics of a population in the next activity, How Will It Land?

How Will It Land?

Goals

- Understand that the probability of some events can be determined only through data collection
- Compute experimental probabilities for samples of increasing size

Materials and Equipment

- A copy of the blackline master "How Will It Land?" for each student or pair of students
- One plastic-foam or paper cup for each student or pair of students
- Four pennies and cellophane tape for each student or pair of students

p. 121

Activity

Distribute the activity sheets to the students, and ask them to read the problem in question 1 and decide what outcomes might be possible. The cup could land upright, upside down, or on its side. Some students may predict that "on its side" is the most likely outcome, whereas others may argue that all three outcomes are equally likely. It is not important that the students explain their reasoning immediately; they will have an opportunity to do so later in the activity.

The students could work either individually or in pairs. It is essential that all the students carry out this experiment under basically the same conditions. In particular, the students should be standing to carry out the simulation, and they will need adequate space in which to flip the cups. The students should attempt to toss their cups in essentially the same manner, flipping them into the air from the same height and letting them fall onto the same surface without interference; failure to flip the cup in a uniform manner may yield biased samples in which one or more of the outcomes is favored. If the students work individually, it may be helpful for them to compare their answers with those of another student before you discuss the activity with the whole class.

Discussion

Ask several students to explain their initial reasoning about which possibility was most likely. Students commonly contend that the cup is more likely to land on its side, reasoning that "there's more surface on the side of the cup than on either the top or the bottom."

Then ask several students to share their results for ten trials and twenty trials of the cup with one penny. Discuss any differences in the students' or pairs' results. If one student contends, "It landed on its side nine times out of ten, so there's a 90 percent chance of its landing that way," ask, "Did anyone get different results?" A classmate may respond, "Mine landed on its side only eight times." Some students may agree with one or the other of the students, some may argue for a different probability, and others may call for additional trials. Survey the class to determine the greatest and lowest ratios for each of the three events. Ask, "How might we find a single value to represent all the data that

Chapter 3: Prediction and the Law of Large Numbers

the class recorded?" Some student may suggest pooling the data. As the pooled sample grows, the experimental probabilities should stabilize and approximate the theoretical (but indeterminate) probability.

If time permits, repeat the discussion, using the data for the trials with four pennies taped in the cup. Ask the students to predict what might happen if ten pennies were taped in the cup. Be alert for how the students are generalizing the patterns that they notice in the data.

The activity How Will It Land? focuses students' attention on how a large sample of simulated data can be used to estimate population characteristics that are impossible to determine through mathematical analysis. Simulations can also be helpful in probabilistic situations for which mathematical analysis is difficult or complex. In such situations, the theoretical probabilities can be estimated reasonably by a simulation that is run an increasing number of times. The next activity, The Long Flight Home, uses the law of large numbers to generate an estimate of theoretical probabilities that may be difficult to determine analytically.

The Long Flight Home

Goals

- Use simulated data to make estimates of theoretical probabilities
- Understand that better inferences can be made from a larger sample than from a smaller sample

Materials and Equipment

- A copy of the blackline master "The Long Flight Home" for each pair of students
- A copy of the blackline master "The Long Flight Home—Extension" for each pair of students who wish to continue their exploration

  This activity sheet is an optional extension. Whether you choose to use it with your students depends on how much experience they have had with permutations and combinations.

- A paper cup for each pair of students
- Two red cubes and two blue cubes for each pair of students
- An overhead projector and transparencies (optional)

pp. 123, 124

Activity

Middle-grades students may have strong intuitions about the fairness of games. For example, they recognize that flipping a coin is an appropriate way to determine which of two players goes first, and they believe that such games as "rock, paper, scissors" are fair. Such intuitions may lead them to believe that it is equally likely that the two people who sit together on the airplane in this activity will be the same gender and that they will be different genders. That is, they may erroneously believe that the event "one male, one female" is just as likely as the event "same gender." This misconception may be a result of the belief that simultaneously drawing two objects from a bag is mathematically different from drawing the two objects in succession without replacement (Falk 1983). In fact, the two methods of drawing two objects are equivalent.

Distribute the first activity sheet to pairs of students, have the students read the problem, and set up the simulation by asking if the two events "same gender" and "different genders" are equally likely. Most students will reason that the two events have the same probability because two people of each gender are involved. Some students may point out that there are two ways to get "same gender"—a pair of blue cubes or a pair of red cubes—and also two ways to get "different genders"—"one blue cube and one red cube" or "one red cube and one blue cube." Because many students will believe that the two events are equally likely, some of them may be indifferent about selecting the part "same" or "different" for the activity or reluctant to make a choice.

Dessart and DeRidder (1999) explain why "rock, paper, scissors" is fair.

Chapter 3: Prediction and the Law of Large Numbers   69

"If students are accustomed to reasoning from and about data, they will understand that discrepancies between predictions and outcomes from a large and representative sample must be taken seriously."
(NCTM 2000, pp. 254–5)

Discussion

After the students have completed the first activity sheet, ask, "Which outcome occurred more often?" Record the results for the class on the board or on transparencies on an overhead projector. Given the relatively long duration of the game (21 trials), it is likely that "different genders" will have occurred more often in most, if not all, of the simulations. Ask the students to explain the reason for the results. Some students may disregard the data and argue that "different" won simply "by luck," reasoning that the results might be different if they took additional draws. The law of large numbers can be employed. Pool the data for the entire class. The pooled data should reflect the theoretical probabilities that are unknown prior to mathematical analysis; in fact, pooling twelve sets of forty-two trials is likely to yield an experimental probability of drawing "different genders" between 62 percent and 71 percent. Use the pooled data to challenge the thinking of your students. Ask, "How can you explain the data?"

Extension

Without analyzing the situation mathematically, the students may erroneously conclude that the likelihood of each event can be judged solely on the basis of the data from an experiment. In reality, although the data may pique students' interest and call into question whether the two events are equally probable, the results do not constitute a proof. Judgments about fairness should be made only after an analysis of the problem—in particular, of the sample space. Questions 1 and 2 on the extension activity sheet will guide the students in such an analysis.

Determining the set of all possible draws of two cubes may prove to be a challenging task for some students. Because the two red cubes are identical (as are the blue cubes), the students may not immediately recognize that drawing a blue cube and a red cube is different from drawing the same blue cube and the other red cube. Have your students try to generate the complete sample space using a list or a tree diagram. Six pairs of equally likely outcomes are possible (see table 3.3).

Table 3.3
Possible Outcomes of Drawing Two Cubes from a Cup with Two Red and Two Blue Cubes

Pair	Same or Different Gender
B_1, B_2	Same gender
B_1, R_1	Different genders
B_1, R_2	Different genders
B_2, R_1	Different genders
B_2, R_2	Different genders
R_1, R_2	Same gender

Thus, P(same gender) = 2/6, or 1/3, and P(different genders) = 4/6, or 2/3. The likelihood that one male and one female will sit together on the flight home is twice the likelihood that both friends will be of the same gender. After determining the sample space in question 1, the students can compare the experimental probabilities of the pooled data (in question 3) with the theoretical probabilities; the proximity of the experimental probabilities to the theoretical probabilities is explained by the law of large numbers.

Aspinwall and Shaw (2000) discuss variations on The Long Flight Home and students' intuitions about probability games.

Conclusion

The first three chapters of this book help students learn some basic ideas about probability. Although these ideas are applied in interesting contexts in these chapters, the main focus is on helping students develop an understanding of the basic ideas. The activities in chapter 4 help students connect these ideas with some aspects of statistics and data analysis. In addition, suggestions are given for the application of probability to other content.

NAVIGATIONS SERIES

GRADES 6–8

NAVIGATING *through* PROBABILITY

Chapter 4
Connecting Probability and Statistics

Important Mathematical Ideas

One important part of mathematics learning is being able to connect and apply concepts to real-world situations, but an equally important part is being able to connect and apply concepts across different areas of mathematics. One of these connections within mathematics is between probability and statistics. *Navigating through Data Analysis in Grades 6–8* (Bright et al. 2003) lays some of the groundwork for statistical ideas, and the first three chapters of this probability book have laid a similar foundation for probability ideas. This chapter presents ways for middle-grades students to explore the connection between probability and statistics by designing and conducting simulations.

A simulation is a procedure for answering questions about a real problem by conducting an experiment that closely resembles the real situation. In order to be useful, a simulation must model the essential characteristics of the situation. However, a simulation is only an approximation of a situation; no simulation will include all the factors that make a real-world situation complex. Indeed, one of the reasons for using a simulation is to eliminate some of the complexity of the real situation so that the relationships among the most important factors can be studied carefully. Using a simulation allows students to explore real-world problems that they cannot study directly because of such obstacles as danger, complexity, or length. Discussing the results of a simulation gives students a deeper understanding of both the simulation process and the situation being simulated.

"Through the grades, students should be able to move from situations for which the probability of an event can readily be determined to situations in which sampling and simulations help them quantify the likelihood of an uncertain outcome."
(NCTM 2000, p. 51)

Probability Model

Each simulation is based on the probabilities of the events in the particular situation. These probabilities form the *probability model* of the simulation. In many real-world situations, we do not know the exact probabilities of the events, so we have to make reasonable assumptions about them. For example, the probability of a girl's being born may not be exactly 1/2, but it is likely to be close to 1/2, so for the sake of simplicity, we assume that the probability is 1/2. The assumption does not seem to be a serious violation of real-world events, so we have confidence that results based on this assumption will give us a sense of the real situation. If the assumptions about a real-world event are incorrect, the results of a simulation of that event may be correspondingly inaccurate. If the assumptions are reasonable, however, the simulation will yield approximations that suffice for many situations.

The results of a simulation will be more precise if it is run repeatedly and a summary of the results is used to predict what might happen in the real world. As noted earlier in this book, probability deals more with long-term trends than with outcomes of individual events. This idea applies equally to the results of simulations. Many cycles of a simulation are needed to obtain the best results.

Genetics is one of the areas in which probability has historically been useful. Brahier (1999) discusses how genetics contexts can be used in the study of probability.

Designing a Simulation

There are five important tasks in designing a simulation:

1. Identify the essential components and assumptions of the problem.
2. Select a random device for the essential components.
3. Define a trial.
4. Conduct a large number of trials and record the information.
5. Use the data to draw conclusions.

Students need to be aware of these tasks so that the simulations they design will be useful. Each of these five tasks will be discussed in the context of the first activity in this chapter, Dixie's Basketball Contest.

Identify Essential Components and Assumptions

First, the problem must be stated clearly so that the necessary information is given and the reason for designing a simulation is clear. The probability model underlying the situation must be specified, but other assumptions may also need to be specified, such as the independence of trials. These assumptions are often not obvious to some students, so the real-world situation may need to be discussed carefully as a prelude to designing a simulation. For example, in Dixie's situation, we assume that the probability of her making any specific free throw is 2/3 and that the free throws are independent. It is possible, of course, that Dixie has a shoulder injury, which might affect both assumptions. But trying to account for all possible circumstances would make designing a simulation impossible. This situation can be analyzed mathematically, without conducting a simulation, but such an analysis requires mathematics knowledge that is not typically part of the middle-grades curriculum. About the only way that most middle-grades students could even address this problem is through the design and execution of a simulation.

Dixie, a basketball player, made 2/3 of her free throws during the season. She has entered a free-throw contest. Each entrant attempts ten free throws. Simulate Dixie's performance in the contest. Use the simulation to determine the approximate probability that Dixie makes at least eight free throws.

Select a Random Device

The conditions of the problem suggest the kinds of random devices that could be used in a simulation. The device must be able to generate chance outcomes with probabilities that match those of the problem. In Dixie's situation, the probability of making a free throw is 2/3, so the probability of not making a free throw is 1/3. The "obvious" device to use is a die or a number cube, but the outcomes on the die can be used in several ways. For example, the values 1, 2, 3, and 4 might represent "making the free throw" and the values 5 and 6 might represent "not making the free throw." Alternatively, noncomposite numbers (i.e., 1, 2, 3, and 5) might represent "making the free throw" and composite numbers (i.e., 4 and 6) might represent "not making the free throw." Indeed, any set of four of the six outcomes can represent "making the free throw" and the other two can represent "not making the free throw."

Define a Trial

One trial of the simulation must be clearly specified. Prior work with probability may lead students to believe that one trial of a simulation must always be only one action, such as one flip of a coin, one spin of a spinner, or one draw from a bag of blocks. Actually, a trial may require several flips, spins, or draws. A trial must model the situation of interest. For Dixie's contest, one trial of the simulation would be ten rolls of a die. Each roll represents one free throw, but the situation being modeled is the complete set of ten free throws, so one trial of the simulation must represent all ten free throws.

Gather Data

The simulation must be executed. Of course, students might decide to conduct only one trial, but as noted earlier, the results will be better if multiple trials are conducted and the data from the trials are summarized in a meaningful way. There is no "magic" number of trials that must be done to yield good data, but as a rule of thumb, twenty-five to thirty trials are appropriate for most of the problems that middle-grades students encounter. The number of trials that are conducted will be influenced by the complexity of the simulation and by the time available for the experiment. For Dixie's situation, each trial requires rolling a die ten times, which takes time and limits the number of trials a single student might be able—or willing—to conduct. When the simulation is complex or when time is short, technology can be very helpful in generating large quantities of data quickly. It is recommended, however, that students conduct at least some trials by hand first, so they understand how the technology is generating the data.

Draw Conclusions

Students frequently believe that they have finished an experiment when they have conducted trials and gathered data. It is equally or perhaps more important, however, that they draw conclusions from the data. The data must often be summarized in some meaningful way, such as by computing an average, making a graph, or simply noting trends in the data. A simulation is useful only if it helps us understand

> "If simulations are used, teachers need to help students understand what the simulation data represent and how they relate to the problem situation."
> (NCTM 2000, p. 254)

the situation that it models. Analyzing the data in the simulation and describing the patterns in the data help us understand the real-world situation. For Dixie's contest, having each student in the class run the simulation five times and then pooling the data are likely to give a nearly accurate approximation of various probabilities, such as the probability that Dixie made at least eight of the free throws. Generating five sets of ten die rolls would not take each student very long, so in only a few minutes of class time, students could conduct more than a hundred trials of the simulation and generate enough data to make inferences that would enhance their understanding of the situation. The students could compare their new understanding of the situation with their original intuitions about it.

Expected Value

The *expected value* of a discrete random variable is the mean of that variable and is computed by first multiplying each value of the variable by the associated probability of that value and then taking the sum of all the products. We do not expect middle-grades students to develop a formal definition of expected value. Rather, they should begin to think about the long-term average of the probability of each outcome weighted by the "payoff" for that outcome. Students may prefer to think of this concept as "what to expect in the long run."

For example, what is the expected value of the numbers generated by tossing one die? Each of the six values—1, 2, 3, 4, 5, and 6—has a probability of 1/6, so the expected value is computed as follows:

$$\begin{aligned} \text{Expected value} &= 1\left(\frac{1}{6}\right) + 2\left(\frac{1}{6}\right) + 3\left(\frac{1}{6}\right) + 4\left(\frac{1}{6}\right) + 5\left(\frac{1}{6}\right) + 6\left(\frac{1}{6}\right) \\ &= \frac{1}{6} + \frac{2}{6} + \frac{3}{6} + \frac{4}{6} + \frac{5}{6} + \frac{6}{6} \\ &= \frac{21}{6} \\ &= 3\frac{1}{2} \end{aligned}$$

Of course, 3 1/2 is not the result of any toss of the die, but over the long term, the average value would be 3 1/2.

Middle-grades students should deal with this concept in the context of averaging many outcomes of an experiment. They should not be expected to understand or use the symbolic representations of this concept. Exploring data from many outcomes will help students appreciate the importance of probability in understanding long-term trends.

What Might Students Already Know about These Ideas?

Students will enter the middle grades with experience in gathering data by tossing coins, spinning spinners, or rolling dice. They will probably have an intuitive understanding that flipping a coin is a device that can be used when two outcomes are equally likely, but they might not understand how to use other devices when the probabilities of events are different from 1/2. The activity Dixie's Basketball Contest will help you determine what your students already understand about designing and using a simulation. You may need to question them about their responses to help them explain their thinking.

Dixie's Basketball Contest

Goal

To assess students'—

- understanding of probabilities of complementary events;
- ability to choose a random device for generating specified probabilities;
- ability to combine simple events to create a simulation of a complex event.

Materials and Equipment

- A copy of the blackline master "Dixie's Basketball Contest" for each student or pair of students
- Dice or spinners
- Chart paper or an overhead projector and transparencies

Activity

Set up this activity by asking the students how difficult it is to make free throws in basketball. If any of the students play on the school's boys' or girls' basketball teams, ask them to talk briefly about how much they have to practice free throws. Ask if making 2/3 of the free throws in a season indicates that a player is particularly skillful. (There may be some disagreement about this point.)

Then ask the students how they have used coins, dice, and spinners to generate data. Their experience in using these devices may vary widely. The point of this discussion is to mention a variety of uses so that the students will have some options to think about as they answer question 2 on the activity sheet.

Distribute a copy of "Dixie's Basketball Contest" to each student or pair of students. Depending on the level of thinking of the students, you may choose to have them work individually or in pairs. It is important, however, that you try to reveal the thinking of individual students during the follow-up discussion.

Discussion

While the students are gathering their data, notice if they are recording all ten tosses of the die (or spins of the spinner) or only the number of baskets made in the ten free throws in each trial. It is better if the students record all the data so that they can check their counts, but it is not necessary to interrupt the students' work to make that point. You can mention it during the discussion. An example of how the table might look is given in figure 4.1.

The students may disagree about the answer to question 3. Their opinions may differ according to their experiences in playing basketball. It is not important that differences of opinion be resolved, although it may be useful for you to know which students believe that Dixie's shots are independent and which students believe that they are dependent.

p. 125

With the applet Spinner on the CD-ROM, students can create a spinner to use in this activity. They could design a three-sector spinner with, for instance, two red sectors representing a basket and one blue sector representing a miss.

For this simulation, a trial will consist of ten tosses of a die or spins of a spinner, since the event being simulated is ten free throws. Each roll of the die or spin represents one of the free throws.

Fig. **4.1.**

Sample data from a simulation of Dixie's free throws

Trial Number	Results of the Trial	Number of Baskets Made
1	2, 1, 1, 5, 2, 4, 6, 5, 3, 1	7
2	1, 3, 2, 4, 5, 1, 2, 3, 6, 1	8
3	2, 2, 6, 3, 4, 1, 2, 3, 6, 6	7
4	1, 2, 4, 4, 1, 5, 6, 3, 1, 4	8
5	5, 2, 4, 4, 1, 4, 5, 3, 3, 5	7

Note: 1, 2, 3, or 4 indicates a basket; 5 or 6 indicates a miss.

You can record the students' answers to question 6 on the overhead projector or on chart paper. Since each student conducts only five trials, the experimental probabilities are likely to show considerable variability. Tally on overhead transparencies or the chart paper the number of times the class's simulations resulted in Dixie's making eight or more baskets and seven or fewer baskets. Ask the students to use these data to answer question 7; this value will be somewhere "in the middle" of the individual estimates.

Conduct a whole-class discussion about question 8. The students may disagree about which estimate is better, so let several of them explain their answers. Help the students categorize the explanations without labeling any "correct" or "incorrect." Some students may say that the class set has more data than any individual set, so the class set is better. Others may say that their method of generating the data was the best method, so their data set is better. Some students may say that they are not sure how other students generated the data, so they cannot be sure that the class data are "good." Since this activity is intended as a preassessment, it is more important for you to understand the students' thinking than it is to reach consensus on all the issues.

Selected Instructional Activities

The remaining activities afford students opportunities to develop their skill at designing simulations and to understand how simulations can be used to study real-world events. No particular order is recommended for the activities. The following activity, Newspaper Route, presents a simple context in which the idea of expected value can be explored.

Newspaper Route

Goals

- Design a simulation
- Determine the expected value experimentally
- Calculate the theoretical expected value

Materials and Equipment

- A copy of the blackline master "Newspaper Route" for each student or pair of students
- Assorted materials to use in the simulation, such as play money, spinners, dice, cards, colored chips, and calculators
- An overhead projector and transparencies

Activity

Distribute the activity sheets, and ask the students to read the problem. Lead a brief discussion about the problem, ensuring that the students understand the payment deal that the customer is proposing. Be particularly careful that the students understand that they must draw the bills out of the bag without looking. You may want to ask for a show of hands from those who think the customer's offer is a good deal. Be sure that the students know what is meant by the word *trial* in question 3. You may want the students to work in pairs so that they can discuss the questions.

Discussion

As the students work, attend to their conversations about the number of $1 bills and how to distinguish one of them from the others. It is important that the students understand that the $1 bills are different. If the bills in the bag were a $1 bill, a $2 bill, a $5 bill, a $20 bill, and a $50 bill (instead of five $1 bills), it would be easy for the students to distinguish them. Discriminating bills of the same denomination, however, is more problematic. One suggestion you can make to the students who are struggling with this task is to suggest that they imagine writing the letters A, B, C, D, and E on the five $1 bills. Then, for example, the $1 bill marked A would be clearly different from the $1 bill marked C. If the students do not distinguish each of the five $1 bills, they very likely will not create the correct probability model for the situation, and consequently their simulation will be inaccurate.

If most of the students are struggling with distinguishing the $1 bills from one another, you might want to lead them through the following experiment: Place play money (or just pieces of paper) in a bag. The five pieces that indicate $1 bills should be labeled "A" through "E." Have the students draw two bills and record both which bills they drew and their value. Record the information in a table on the overhead projector. After six or eight pairs of bills have been drawn, ask the students to organize a table that shows all the possible combinations of two of the five bills.

p. 127

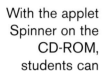

With the applet Spinner on the CD-ROM, students can create a spinner to use in this activity. They could design a six-sector spinner with, for instance, one red sector representing $11 and five blue sectors representing $2.

The chart in figure 4.2 shows the fifteen possible outcomes of drawing two bills out of the bag. There are five ways to get $11 and ten ways to get $2. A common error committed by students is to list $2 and $11 as the two outcomes and assume that they are equally likely. It is acceptable to think of $2 and $11 as the two outcomes but only if it is made very clear that the probabilities of these two outcomes are different.

$1	$1	$1	$1	$1	$10	Total Value
X	X					$2
X		X				$2
X			X			$2
X				X		$2
X					X	$11
	X	X				$2
	X		X			$2
	X			X		$2
	X				X	$11
		X	X			$2
		X		X		$2
		X			X	$11
			X	X		$2
			X		X	$11
				X	X	$11

Fig. **4.2.**

A chart of the sample space for drawing two bills from the bag

After the students have completed the activity sheet, ask two or three of them to explain how they simulated the customer's offer. Listen carefully to their explanations of the random device they chose and how they used the device in the simulation. For example, some students might draw play money out of a bag. Others might use two dice, with the value 6 representing the $10 bill and each of the other values representing the $1 bills. If they use dice, the students must be very explicit that doubles—for example, (1, 1) or (4, 4)—must be eliminated, since no bill can be drawn twice from the bag. No matter what random device the students choose to use, it is important that they be clear about how each outcome from the device is associated with each outcome of the situation.

Compile all the answers for question 4, and display them on the overhead projector. Ask the students if drawing $2 and drawing $11 are

A graphing calculator could be used instead of a die to generate random numbers for the simulation. The first pair of random numbers displayed would have to be discarded, since bill 1 cannot be drawn twice.

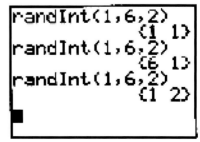

Chapter 4: Connecting Probability and Statistics

equally likely. The data should be convincing that the probability of drawing $2 is approximately twice as great as the probability of drawing $11. It is not essential that all the students be able to give the precise probabilities of 2/3 and 1/3 for $2 and $11, respectively, but some students may be able to explain the probabilities. Those who can may argue from data they have organized in a chart similar to that in figure 4.2.

Ask the students to tell you the mean amount of money drawn per week from the bag (question 5), and compile these answers on the overhead projector. The answers will exhibit variability, but they should cluster around $5 per week, since $5 is the expected value for this situation. The original payment option would also net $5 per week. Depending on the level of your students' thinking, you may choose to show the actual expected value, which is based on the probability model or probability distribution of the situation. The probability of getting $2 is 2/3, and the probability of getting $11 is 1/3. The expected value is

$$\$2\left(\frac{2}{3}\right) + \$11\left(\frac{1}{3}\right) = \frac{\$4}{3} + \frac{\$11}{3}$$
$$= \frac{\$15}{3}$$
$$= \$5.$$

The expected value can also be computed for the fifteen different possible draws by recognizing that $2 would happen ten times and $11 would happen five times. The average is

$$\frac{[\$2(10) + \$11(5)]}{15} = \frac{(\$20 + \$55)}{15}$$
$$= \frac{\$75}{15}$$
$$= \$5.$$

Over the long run, the two methods of payment can be expected to generate the same amount of income for the carrier.

Ask the students whether the customer's offer is a good deal for the carrier (question 7). Some students may argue that it is not a good deal. Their arguments may be based on various factors. Some students may say that the mean they got in question 5 is less than $5, so the carrier will lose money. This argument would reflect only their particular data without acknowledging long-term trends. Some students may say that the probability of getting less than $5 is greater than the probability of getting more than $5. This argument contains some elements of an argument based on the expected value, but it is incomplete. If, for example, the probability of drawing $2 were .51 and the probability of drawing $11 were .49, the argument would still apply, but the expected value of this option would be $6.41. In this hypothetical case, the carrier should definitely accept the alternative payment plan.

Some students may recognize that the alternative payment plan will generate the same income over the long term as the original payment plan. These students will likely have an intuitive understanding of the idea of expected value, although they may not be able to explain the concept clearly. The carrier's decision should be based not on the outcome for a single week but on the long-term average—that is, the expected value. If the distribution of the bills in the alternative plan were different and resulted in long-term average payments that were great enough, drawing money out of the bag would be worth the risk.

Extension

As an extension, the students could consider what would happen if there were different distributions of bills: for example, two $10 bills and four $1 bills, or one $10 bill and eight $1 bills. Which set of bills in the bag is more favorable to the carrier? Which is more favorable to the customer?

The next activity is How Black Is a Zebra? Its effectiveness would be enhanced by students' use of such technology as a graphing calculator.

How Black Is a Zebra?

Goal

- Explore the use of sampling as a means of answering a question
- Use a random-number generator to determine the coordinates of random points
- Explore the importance of sample size when making a prediction or estimate

Materials and Equipment

- A copy of the blackline master "How Black Is a Zebra?" for each pair of students.
- A random-number generator, such as a graphing calculator or a table of random numbers
- An overhead projector and transparencies

Activity

Distribute the activity sheets, which include a drawing of a zebra. Ask the students to estimate what percent of the zebra is black. Record their guesses on a transparency on the overhead projector, but do not try to reach a consensus on an estimate.

Have the students work in pairs to complete the activity sheets. Be sure that the students know how to use the random-number generator to generate pairs of numbers that will serve as the coordinates of points. Also, make sure that the students understand that if a point does *not* land on the zebra, that point should be discarded and *not* recorded in the table on the activity sheet. Explain that the grid that has been superimposed on the picture will help them determine if a point falls on the zebra and where it appears.

Discussion

Observe the students as they record the data points to be sure that they are including only the points that actually fall on the zebra. After the students have finished, ask a few of them to share their findings for the first ten data points. Listen carefully to their explanations of whether they think their estimates were accurate. They should be cautious about the accuracy of an estimate that is based on so few points.

Ask one or two students to discuss their twenty-point samples and to explain whether they are more or less confident about the accuracy of the estimate based on twenty points than they were about the estimate based on ten points. The students should express greater confidence in the later estimates because the number of points on which they based them is greater. But twenty points still do not make a very large sample, so the students should register some hesitancy about the estimates.

Then compile all the data from the class by determining the total number of black hits generated by the class and dividing that total by the number of pairs of students. Ask the students to generate a class estimate of the percent of the zebra that is black. (They should do so by dividing the class average of black hits by 30.) Then discuss the

p. 128

Of the three pairs of randomly generated numbers on the calculator display—(12, 11), (22, 17), and (20, 4)—only the first and last would be recorded as data points, since they both fall on the zebra. The point (22, 17) does not fall on the zebra, so it would not be recorded in the table.

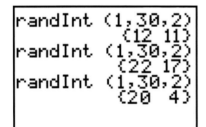

Suppose that ten pairs of students found that the following percents of the first ten points fell on black: 40%, 80%, 90%, 10%, 70%, 30%, 70%, 100%, 30%, and 40%. The great variability in the data might make some students wonder if they were using different pictures. However, ten points constitute a very small sample, which is not likely to be very representative of the population of all the points on the zebra picture.

differences between the predictions from the pairs' small samples and the larger, pooled sample. The students will probably express considerably more confidence in the class estimate. This discussion should give you insight into the students' understanding of making estimates and predictions on the basis of gathered data.

Conclusion

The activities in this book are designed to help students develop an understanding of some important probability concepts and then connect those concepts with aspects of data analysis. A notion of probabilities as relative frequencies and an understanding of the importance of identifying long-term trends are essential to developing good reasoning about probability and data analysis. Students also need to understand that they can make predictions about a population by examining a representative sample from that population and that a random sample is acceptable for that purpose. The ideas presented in this book lay the foundation for the more sophisticated reasoning that is expected of students in high school.

A computer program, Scion Image (Scion Corp., n.d.), was used to determine that the zebra in the picture is 53.77 percent black. Students who are interested in knowing the range of actual percents of black on real zebras could contact a biologist or a zoo.

NAVIGATIONS SERIES

GRADES 6–8

NAVIGATING through PROBABILITY

Looking Back and Looking Ahead

In the middle grades, work in probability begins with understanding probability as a ratio of successful outcomes to the total number of outcomes. Students then begin to explore compound events and to learn about independent events, and these activities prepare them to understand probability distributions and the law of large numbers. Along the way, students develop increasingly sophisticated language and representations that help them not only communicate their reasoning but also engage in probabilistic reasoning. Students learn to use tables and graphs to communicate information about probabilities, so it is important to connect probability instruction with instruction in making and interpreting graphs.

Many of the obvious contexts for applying probability concepts are games that use random devices, such as dice or spinners. Games are motivating for students, but the use of games requires teachers to find ways to connect what students learn in game-playing environments with what they encounter in more real-world settings. It is especially important that students be allowed to confront their often faulty intuitions in settings that are nonthreatening. Probability intuitions are often deeply ingrained; such intuitions will change only through repeated explorations of ideas in different contexts.

Much of the reasoning used by middle-grades students in probability situations is informal. More-advanced topics, such as permutations and combinations, are not studied until high school or beyond, when more-formal mathematical tools are available to students. However, it is important for middle-grades students to begin to understand the need for precision in reasoning and in communicating reasoning. One significant challenge is to guide students in choosing strategies carefully and then in communicating clearly how the chosen strategies support their conclusions in given contexts.

Technology can be useful in middle-grades students' explorations of probability, but it cannot be expected to take the place of direct, hands-on work (e.g., tossing dice). Graphing software and graphing calculators can help students make different kinds of graphs of data from probability experiments quickly and easily. Technology can also be used to generate large numbers of trials, which might be especially important for understanding the law of large numbers. Prior to using computer simulations, however, students should generate trials with hands-on materials so that they can experience how an experiment really works.

Middle-grades teachers must help students make the transition from the relatively unsophisticated probability reasoning typical in elementary school to the more-formal reasoning that is developed in high school. Providing a wide range of experiences with different kinds of random devices in many different contexts is necessary to this process. Students need to analyze situations abstractly but also to conduct simulations to generate empirical probabilities so that they can understand how empirical and experimental probability are connected and how the two concepts reinforce each other. Dealing with all these ideas requires careful planning and sequencing of instruction. Connecting probability with other content—for example, data analysis and science—can make effective use of the available instructional time and help students develop connections both within mathematics and between mathematics and other subjects.

Probability is not only a motivating topic for students but also an increasingly important part of mathematics. It is important that middle-grades students develop a firm understanding of the basic ideas of probability and then learn how these ideas can be applied in different situations.

Navigating through Probability

Grades 6–8

Appendix
Blackline Masters and Solutions

Who Will Win?

Name _____

Suppose that the spinner below is used in a simple game for three players. These are the rules of the game:

- Each player chooses a color, and the players take turns spinning the spinner.
- On each spin, the person whose color is spun gets 1 point.
- The winner is the first person to accumulate 10 points.

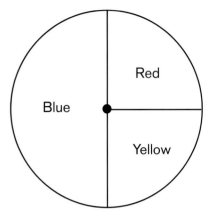

1. What are the possible outcomes on each spin? _____
   _____

2. Is the game fair? _____ Explain your answer. _____
   _____
   _____

Suppose that the game rules remain the same but that the spinner on the right is used.

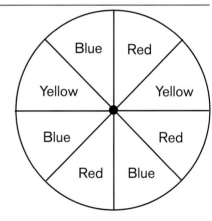

3. What are the possible outcomes on each spin? _____
   _____

4. Is the game fair? _____ Explain your answer. _____
   _____
   _____

5. On the back of this page, draw three different spinners, each of which would make the game fair.

6. How are your spinners the same? _____
   _____
   _____

   How are they different? _____
   _____
   _____

90   Navigating through Probability in Grades 6–8

Two Dice

Name _____

1. If you rolled two dice at the same time and summed the numbers, what sums could you get? _____
   _____

   Would they be equally likely? _____ If not, which one or ones would be most likely? _____
   _____ Least likely? _____ Explain your answers. _____
   _____
   _____

2. Suppose that one die is red and the other is green. Make a chart to show all the pairs of numbers that could occur if you rolled the two dice. What sums would those pairs generate?

3. If you rolled one die, recorded the result, tossed the same die again, recorded the result, and then summed the two numbers, what sums could you get? _____
   _____

   Would they be equally likely? _____ If not, which one or ones would be most likely? _____
   _____ Least likely? _____ Explain your answers. _____
   _____
   _____

4. Show all the pairs of numbers that could result if you rolled the same die twice. What sums would those pairs generate?

Navigating through Probability in Grades 6–8

Two Dice (continued)

Name _____

5. How are the pairs of numbers you showed in questions 2 and 4 alike? _____

    How are they different? _____

6. Suppose that you—

    • rolled one die, doubled the value of the result, and recorded that number, and

    • rolled the die again and added the number to the number you just recorded.

    What values could result?

    Are these values equally likely? _____ If not, which one or ones are most likely? _____

    Least likely? _____ Explain. _____

    _____

    _____

Pets

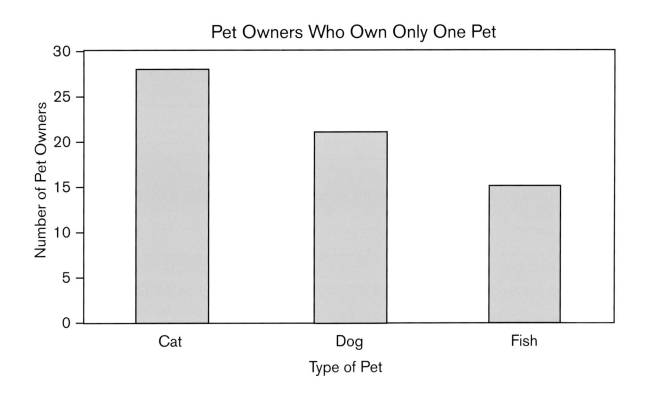

More Often and Most Often

Name _____

Suppose that you have a bag that contains six candies: three mints, two butterscotch drops, and one caramel, with the candies thoroughly mixed. Decide which of the following statements are true and which are not true. Explain each of your answers.

1. If you draw out one candy without looking, the probability of drawing a mint is greater than the probability of drawing a candy that is not a mint. _____
   _____
   _____

2. If you draw out one candy without looking, the candy is more likely to be a mint than any other kind.
   _____
   _____

3. If you repeated the process of drawing out a candy and putting it back in the bag many times, the kind of candy drawn out is likely to be mint most of the time. _____
   _____
   _____

4. If you repeated the process of drawing out a candy and putting it back in the bag many times, the candies drawn out are likely to be "mint or caramel" most of the time. _____
   _____
   _____

Ratios

Names _____

_____

1. Roll two dice, and find the sum of the values indicated on the top faces. Decide whether the sum is prime or not prime. Put an *X* in the appropriate column. Then compute the ratio of "number of prime sums" to "cumulative number of rolls," and enter the ratio in the chart. Repeat the process twenty times.

Roll	Numbers Rolled and Sum	Prime Sum?	Not a Prime Sum?	Ratio of "Number of Prime Sums" to "Cumulative Number of Rolls"
1				
2				
3				
4				
5				
6				
7				
8				
9				
10				
11				
12				
13				
14				
15				
16				
17				
18				
19				
20				

Navigating through Probability in Grades 6–8

Ratios (continued)

Names _____
_____

2. Graph the data on the following axes:

Ratio of Number of Prime Sums to Number of Rolls

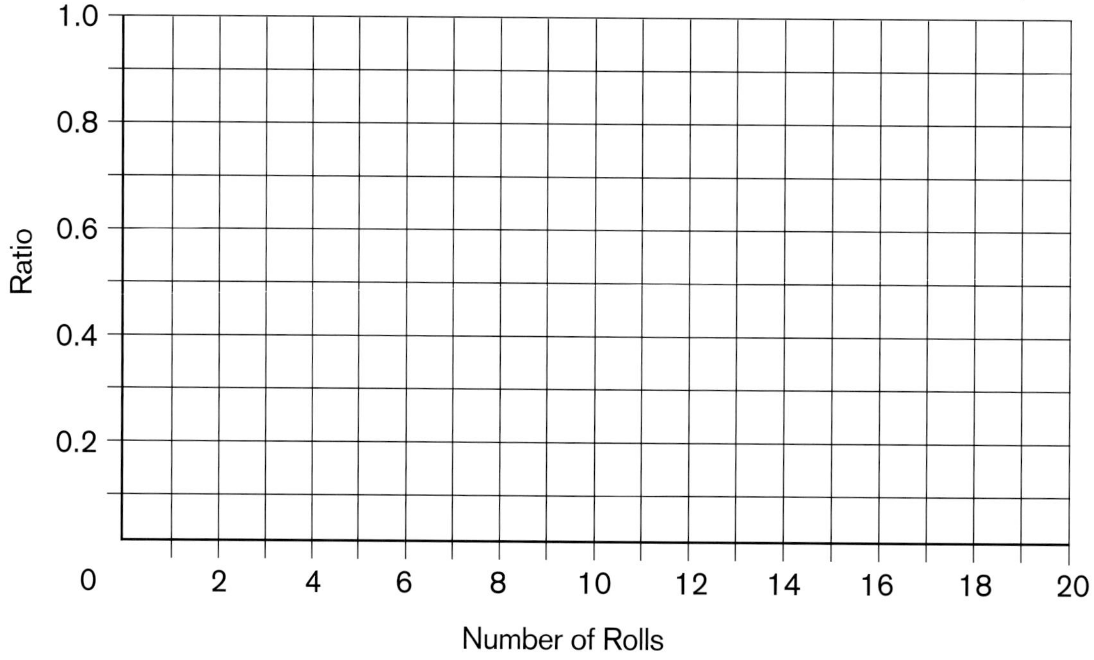

[Graph with y-axis labeled "Ratio" from 0 to 1.0 in increments of 0.2, and x-axis labeled "Number of Rolls" from 0 to 20 in increments of 2.]

3. What patterns do you see in the graph? _____
_____
_____

   How does the graph of the first ten rolls differ from the graph as the number of rolls increases?
_____
_____

4. If you rolled the dice many times, about what ratio of sums do you think would be prime? _____
   Explain your answer. _____
_____
_____

Navigating through Probability in Grades 6–8

Ratios (continued)

Names _____

_____

5. Roll two dice, and multiply the values indicated on the top faces each time. Observe whether the product is a two-digit number or not. Put an *X* in the appropriate column. Then compute the ratio of "number of two-digit products" to "cumulative number of rolls," and enter the ratio in the chart. Repeat the process twenty times.

Roll	Numbers Rolled and Product	Two-Digit Product?	Not a Two-Digit Product?	Ratio of "Number of Two-Digit Products" to "Cumulative Number of Rolls"
1				
2				
3				
4				
5				
6				
7				
8				
9				
10				
11				
12				
13				
14				
15				
16				
17				
18				
19				
20				

Navigating through Probability in Grades 6–8

Ratios (continued)

Names _____
_____

6. Graph the data on the following axes:

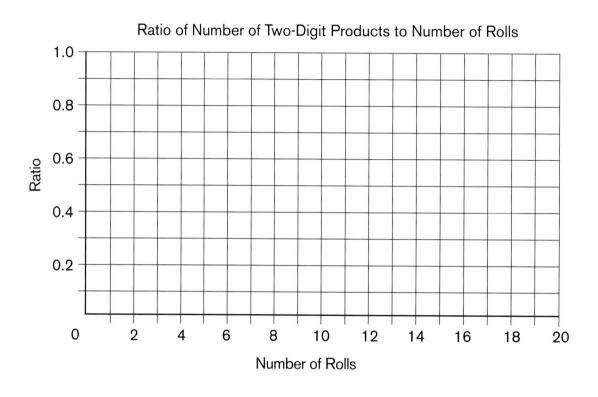

7. What patterns do you see in the graph? _____
_____
_____

How does the graph of the first ten rolls differ from the graph as the number of rolls increases?
_____
_____

8. If you rolled the dice many times, about what ratio of products do you think would be two-digit numbers? _____ Explain your answer. _____
_____
_____

Fair Spinners

Name _____

Chris and Pat work in the quality-control section of a company that makes spinners. Their job is to be sure that the spinners are fair. The process is to select a spinner at random from the production line, spin it a few dozen times, and record what colors the spinner lands on. On Thursday afternoon, Chris and Pat each choose a spinner like the one below from the production line. Their data appear below the spinner. Pat spins faster than Chris, so they do not record the same number of spins.

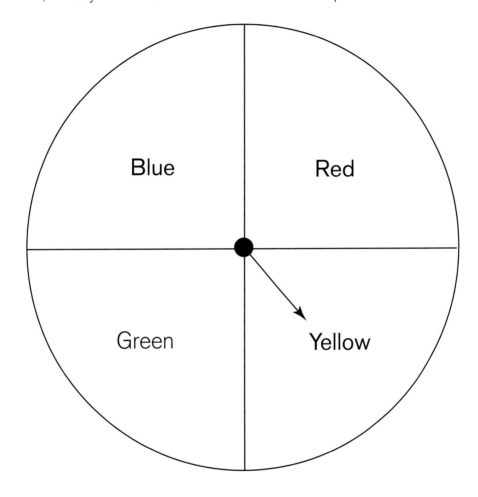

	Color			
	Red	Blue	Green	Yellow
Chris's spins	8	11	11	10
Pat's spins	16	14	18	12

Navigating through Probability in Grades 6–8

Fair Spinners (continued)

Name _____

1. Chris thinks that a graph of these data can help them decide if the spinners are fair. Make a double-bar graph to show these data.

2. What does this graph tell you about whether the spinners are fair? _____
_____
_____
_____

Fair Spinners (continued)

Name _____

3. Chris and Pat decide to compare the two spinners. Pat suggests that a relative-frequency graph can help them compare the data. Make a double-bar graph to show the relative frequencies.

4. What does the relative-frequency graph tell you about which spinner is more fair? _____
   _____
   _____
   _____

5. How are your answers to questions 2 and 4 the same? _____
   _____
   _____

   How are they different? _____
   _____
   _____

Navigating through Probability in Grades 6–8

Dice Differences

Names _____
      _____

This game is for two players.

Rules
1. Decide who will be player A and who will be player B.
2. Decide who goes first, and then alternate turns.
3. On your turn, do the following:
   - Roll two dice.
   - If the values are different, subtract the lesser value from the greater value.
   - Record the difference in the spaces below. (See the sample game.)
   - If the values are the same, enter a zero.
4. Play at least three games.

Scoring
- Player A scores a point if the difference is 0, 1, or 2.
- Player B scores a point if the difference is 3, 4, or 5.
- The winner of a game is the first player to score 10 points.

Sample game. _0_, _2_, _1_, _2_, _3_, _3_, _1_, _0_, _1_, _4_,
              _1_, _2_, _1_, _3_, ___, ___, ___, ___, ___, ___

   Player A's tally of points: |||| ||||
   Player B's tally of points: ||||

Game 1.   ____, ____, ____, ____, ____, ____, ____, ____, ____,
              ____, ____, ____, ____, ____, ____, ____, ____, ____

   Player A's tally of points: _____
   Player B's tally of points: _____

Game 2.   ____, ____, ____, ____, ____, ____, ____, ____, ____,
              ____, ____, ____, ____, ____, ____, ____, ____, ____

   Player A's tally of points: _____
   Player B's tally of points: _____

Game 3.   ____, ____, ____, ____, ____, ____, ____, ____, ____,
              ____, ____, ____, ____, ____, ____, ____, ____, ____

   Player A's tally of points: _____
   Player B's tally of points: _____

Dice Differences (continued)

Name _____

Game 4. _____, _____, _____, _____, _____, _____, _____, _____, _____, _____,
_____, _____, _____, _____, _____, _____, _____, _____, _____, _____

    Player A's tally of points: _____

    Player B's tally of points: _____

Game 5. _____, _____, _____, _____, _____, _____, _____, _____, _____, _____,
_____, _____, _____, _____, _____, _____, _____, _____, _____, _____

    Player A's tally of points: _____

    Player B's tally of points: _____

For each game, record the number of times each difference occurred:

	\	\	Difference	\	\	\
	0	1	2	3	4	5
Game 1	__	__	__	__	__	__
Game 2	__	__	__	__	__	__
Game 3	__	__	__	__	__	__
Game 4	__	__	__	__	__	__
Game 5	__	__	__	__	__	__

1. How many games did player A win? _____

2. How many games did player B win? _____

3. Is this a fair game? _____ Explain your answer. Use the data above to help you.

_____

_____

Test Guessing

Names _____
_____
_____

Suppose that you have to take a true-or-false test with three questions and you have forgotten to study. You take the test, but you have to guess on each question.

1. Since there are only two choices for each question (true or false), what is the probability that you will guess the correct answer for the first question? _____ For the second question? _____ For the third question? _____

2. Using C for a correct guess and W for a wrong guess, list all the possible outcomes of answering the three questions on the test. For example, you would record CCC to indicate the possibility of getting all three questions correct. *Hint:* Eight outcomes are possible.

3. If you are truly guessing, what is the probability associated with each of the eight outcomes? _____
   _____
   Describe two ways to explain your answer. _____
   _____
   _____

4. If 70 percent is the lowest passing grade, what is the probability that you will pass the test by guessing? _____

5. Repeat the analysis (questions 1–4) for a true-or-false test of five questions.

6. If time permits, repeat the analysis (questions 1–4) for a three-question multiple-choice test with three options for each answer.

Number Golf: One Die

Name _____

This game is for two to four players, each with his or her own activity sheet.

Rules

1. If there are two players, the shorter one plays first, and then the players take turns. If there are more than two players, the shortest person plays first. Then play continues clockwise.
2. Each player begins the first hole with a cumulative number of 0.
3. On his or her turn, the player rolls the die.
4. A player may add the number on the die to his or her current cumulative number or subtract it from the current cumulative number.
5. Each player records the number on the die and his or her new cumulative number in the spaces provided on the activity sheet.
6. A player's objective for each hole is for the cumulative number to equal the goal number for that hole.
7. For each hole, a player's score is the number of times she or he rolls the die to reach the goal. If a player rolls the die ten times without reaching the goal, she or he stops play on that hole and records 10 as the score for that hole.
8. When a player begins a new hole, her or his cumulative number (the "tee-off number") begins with the goal number for the previous hole.
9. The winner is the player with the lowest total score for three holes.

Hole 1 (Goal = 15)		Hole 2 (Goal = 25)		Hole 3 (Goal = 29)	
Value on Die	Cumulative Number	Value on Die	Cumulative Number	Value on Die	Cumulative Number
Tee-Off Number	0	Tee-Off Number	15	Tee-Off Number	25
_____	_____	_____	_____	_____	_____
_____	_____	_____	_____	_____	_____
_____	_____	_____	_____	_____	_____
_____	_____	_____	_____	_____	_____
_____	_____	_____	_____	_____	_____
_____	_____	_____	_____	_____	_____
_____	_____	_____	_____	_____	_____
_____	_____	_____	_____	_____	_____
_____	_____	_____	_____	_____	_____
_____	_____	_____	_____	_____	_____

Goals

Hole 1: **15**

Hole 2: **25**

Hole 3: **29**

Scores

Hole 1: _____

Hole 2: _____

Hole 3: _____

Total score for three holes: _____

Navigating through Probability in Grades 6–8

Number Golf: Two Dice

Name _____

This game is for two to four players, each with his or her own activity sheet.

Rules
1. If there are two players, the taller one plays first. If there are more than two players, the tallest person plays first. Then play continues counterclockwise.
2. Each player begins the first hole with a cumulative number of 0.
3. On his or her turn, a player rolls the dice and adds the two values on the top faces.
4. The player may add the sum to his or her current cumulative number or subtract the sum from the current cumulative number.
5. Each player records the sum for the roll and the new cumulative sum in the spaces provided on the activity sheet.
6. A player's objective for each hole is for the cumulative number to equal the goal for that hole.
7. For each hole, a player's score is the number of times she or he rolls the dice to reach the goal.
8. If a player rolls the dice ten times without reaching the goal, she or he stops play on that hole and records 10 as the score for that hole.
9. When a player begins a new hole, her or his cumulative number (the "tee-off number") begins with the goal for the previous hole.
10. The winner is the player with the lowest total score for three holes.

Hole 1 (Goal = 18)		Hole 2 (Goal = 28)		Hole 3 (Goal = 33)	
Sum of Values on Dice	Cumulative Number	Sum of Values on Dice	Cumulative Number	Sum of Values on Dice	Cumulative Number
Tee-Off Number	0	Tee-Off Number	18	Tee-Off Number	28
_____	_____	_____	_____	_____	_____
_____	_____	_____	_____	_____	_____
_____	_____	_____	_____	_____	_____
_____	_____	_____	_____	_____	_____
_____	_____	_____	_____	_____	_____
_____	_____	_____	_____	_____	_____
_____	_____	_____	_____	_____	_____
_____	_____	_____	_____	_____	_____
_____	_____	_____	_____	_____	_____
_____	_____	_____	_____	_____	_____

Goals

Hole 1: **18**

Hole 2: **28**

Hole 3: **33**

Scores

Hole 1: _____

Hole 2: _____

Hole 3: _____

Total score for three holes:
_____

First Head

Name _____

Exploration 1

1. Use any fair coin. Toss it repeatedly, recording the outcome of each toss as H or T, until the first head appears. Then record the number of the toss on which the first H appears. Repeat this process twenty times.

Trial	String of Hs and Ts	Number of Tosses to Get First Head
Sample	T T T H	4
1	_____	____
2	_____	____
3	_____	____
4	_____	____
5	_____	____
6	_____	____
7	_____	____
8	_____	____
9	_____	____
10	_____	____
11	_____	____
12	_____	____
13	_____	____
14	_____	____
15	_____	____
16	_____	____
17	_____	____
18	_____	____
19	_____	____
20	_____	____

2. Estimate the probability of each event by computing relative frequencies:

Event	Estimate of Probability	Event	Estimate of Probability
First H on toss 1	_____	First H on toss 7	_____
First H on toss 2	_____	First H on toss 8	_____
First H on toss 3	_____	First H on toss 9	_____
First H on toss 4	_____	First H on toss 10	_____
First H on toss 5	_____	No H in 10 tosses	_____
First H on toss 6	_____		

3. What surprised you about the number of times you tossed a coin before you got a head? _____
_____
_____
_____

Navigating through Probability in Grades 6–8

First Head (continued)

Name _____

Exploration 2

1. Tape a small wad of paper to the tail side of the coin. Repeat the trials.

Trial	String of Hs and Ts	Number of Tosses to Get First Head
Sample	T T T T T H	6
1	_____	____
2	_____	____
3	_____	____
4	_____	____
5	_____	____
6	_____	____
7	_____	____
8	_____	____
9	_____	____
10	_____	____
11	_____	____
12	_____	____
13	_____	____
14	_____	____
15	_____	____
16	_____	____
17	_____	____
18	_____	____
19	_____	____
20	_____	____

2. Estimate the probability of each event by computing relative frequencies:

Event	Estimate of Probability	Event	Estimate of Probability
First H on toss 1	_____	First H on toss 7	_____
First H on toss 2	_____	First H on toss 8	_____
First H on toss 3	_____	First H on toss 9	_____
First H on toss 4	_____	First H on toss 10	_____
First H on toss 5	_____	No H in 10 tosses	_____
First H on toss 6	_____		

3. How were the results of these two explorations the same? _____
_____
_____

  How were they different? _____
_____
_____

Strings of Heads

Names _____

_____

Exploration 1

1. Use any fair coin. Toss it ten times. Record the longest string of heads in the ten tosses. For example, if the tosses are H H T H H H T T T T, the longest string of heads is 3. Repeat this process twenty times, making ten tosses each time.

Trial	Tosses	Longest String of Heads
Sample	H H T H H H T T T T	3
1	_____	_____
2	_____	_____
3	_____	_____
4	_____	_____
5	_____	_____
6	_____	_____
7	_____	_____
8	_____	_____
9	_____	_____
10	_____	_____
11	_____	_____
12	_____	_____
13	_____	_____
14	_____	_____
15	_____	_____
16	_____	_____
17	_____	_____
18	_____	_____
19	_____	_____
20	_____	_____

2. Estimate the probability of each event by computing relative frequencies:

Event	Estimate of Probability	Event	Estimate of Probability
Length of longest string = 0	_____	Length of longest string = 5	_____
Length of longest string = 1	_____	Length of longest string = 6	_____
Length of longest string = 2	_____	Length of longest string = 7	_____
Length of longest string = 3	_____	Length of longest string = 8	_____
Length of longest string = 4	_____	Length of longest string = 9	_____
		Length of longest string = 10	_____

Navigating through Probability in Grades 6–8

Strings of Heads (continued)

Names _____
_____

3. Are you surprised by any of your results? _____ What surprised you? _____
   _____
   _____

4. Do you think the length of the longest string of heads would be different if you tossed the coin twenty times in each trial? _____
   Explain. _____
   _____
   _____

Exploration 2

1. Use any fair coin. Toss it twenty times. Record the longest string of heads in the twenty tosses. Repeat this process twenty times, making twenty tosses each time.

Trial	Tosses	Longest String of Heads
1		
2		
3		
4		
5		
6		
7		
9		
10		
11		
12		
13		
14		
15		
16		
17		
18		
19		
20		

Strings of Heads (continued)

Names _____

_____

2. Estimate the probability of each event by computing relative frequencies:

Event	Estimate of Probability
Length of longest string = 0	_____
Length of longest string = 1	_____
Length of longest string = 2	_____
Length of longest string = 3	_____
Length of longest string = 4	_____
Length of longest string = 5	_____
Length of longest string = 6	_____
Length of longest string = 7	_____
Length of longest string = 8	_____
Length of longest string = 9	_____
Length of longest string =10	_____
Length of longest string = 11	_____
Length of longest string = 12	_____
Length of longest string = 13	_____
Length of longest string = 14	_____
Length of longest string = 15	_____
Length of longest string = 16	_____
Length of longest string = 17	_____
Length of longest string = 18	_____
Length of longest string = 19	_____
Length of longest string = 20	_____

3. How were the results of the twenty-toss exploration similar to the results of the ten-toss exploration?
_____
_____
_____

How were they different? _____
_____
_____
_____

Strings of Heads (continued)

Names _____

_____

Exploration 3

1. Get the results of all the games that your favorite baseball or basketball team played last year. List each game as a win (W) or a loss (L). Record the longest string of wins in each set of ten games during the season (e.g., games 1–10, games 11–20).

Games	Wins/Losses	Longest String of Wins
1–10	_____	_____
11–20	_____	_____
21–30	_____	_____
31–40	_____	_____
41–50	_____	_____
51–60	_____	_____
61–70	_____	_____
71–80	_____	_____
81–90	_____	_____
91–100	_____	_____
101–110	_____	_____
111–120	_____	_____
121–130	_____	_____
131–140	_____	_____
141–150	_____	_____
151–160	_____	_____

2. Estimate the probability of each event by computing relative frequencies:

Event	Estimate of Probability	Event	Estimate of Probability
Length of longest string = 0	_____	Length of longest string = 6	_____
Length of longest string = 1	_____	Length of longest string = 7	_____
Length of longest string = 2	_____	Length of longest string = 8	_____
Length of longest string = 3	_____	Length of longest string = 9	_____
Length of longest string = 4	_____	Length of longest string = 10	_____
Length of longest string = 5	_____		

3. Do the data look like the data for a coin toss? _____ Explain. _____

    _____

    _____

4. What difference between the sports situation and the coin-toss situation might account for the differences you observed in the results? _____

    _____

    _____

    _____

Racing Game

Name _____

This game is a race among eleven cars, numbered 2 through 12. The numbers of the cars correspond to the possible sums that result when two dice are rolled. These are the rules:

1. Roll two dice, and add the values on the top faces.
2. In the score chart below, place an *X* in the column for the car whose number corresponds to the sum of the values that were rolled. Begin at the bottom of the chart, and place the *X*s in successive boxes.
3. In the second column, record the number of the car that first reaches each of the mileposts.
4. The winner is the first car to reach milepost 10.

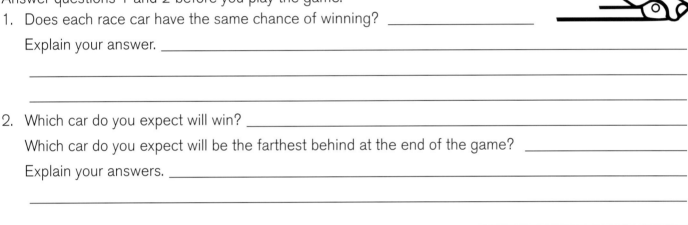

Answer questions 1 and 2 before you play the game.

1. Does each race car have the same chance of winning? _____

   Explain your answer. _____

   _____

   _____

2. Which car do you expect will win? _____

   Which car do you expect will be the farthest behind at the end of the game? _____

   Explain your answers. _____

   _____

   _____

Score Chart

Milepost	First Car to Reach Milepost	Car										
		2	3	4	5	6	7	8	9	10	11	12
10												
9												
8												
7												
6												
5												
4												
3												
2												
1												

Navigating through Probability in Grades 6–8

Racing Game (continued)

Name _____

3. Which car won? _____ Which car was the farthest behind? (More than one answer is possible.) _____

4. Pool the results for your class. Which car won most often? _____
   Which car was the farthest behind at the end of the game? (More than one answer is possible.) _____
   _____

5. If you repeated the race, which car do you think would win? _____
   Explain your answer. _____
   _____
   _____

Pooled Racing-Game Data

Name _____

Car 2	Car 3	Car 4	Car 5	Car 6	Car 7	Car 8	Car 9	Car 10	Car 11	Car 12

Two Hospitals

Name _____

Two hospitals keep track of the gender of the babies born each day. Hospital A is a large urban medical center. Hospital B is a small regional facility. Many more babies are born each day in Hospital A than in Hospital B. Assume that for each birth, the probability that the baby is male is .5 and the probability that the baby is female is .5.

1. Which of the following statements is more likely to be correct? _____

    A. At least 20 out of 25 of the babies born in Hospital A are female.
    B. At least 4 out of 5 of the babies born in Hospital B are female.

    Or are these two events equally likely to occur?

    Explain your choice. _____
    _____
    _____
    _____

2. In both events, at least 80% of the births must be female, but the total number of births in the two events is different. Is it more likely that 80% of the births will be female in a small sample or in a large sample, or is the likelihood equal for large and small samples? _____
    Explain your answer. _____
    _____
    _____
    _____

Let's do an experiment about 5 births. To get ready, answer questions 3 and 4.

3. How many female births do you think will occur in one set of 5 trials? _____ Explain your answer. _____
    _____
    _____
    _____

4. What would be an unusual number of female births in one set of 5 trials? _____ Explain your answer. _____
    _____
    _____

Now conduct the experiment:
- Use a two-color chip to simulate the births of 5 children.
- Let one color represent boys and the other color represent girls.
- Toss the chip 5 times, and record the number of boys and girls born.

5. How many female babies were born in 5 births? _____ Did the result agree with your prediction? _____

Two Hospitals (continued)

Name _____

Let's do an experiment about 25 births. To get ready, answer questions 6 and 7.

6. How many female births do you think will occur in one set of 25 trials? _____
   Explain your answer. _____
   _____
   _____

7. What would be an unusual number of female births in one set of 25 trials? _____
   Explain your answer. _____
   _____
   _____

Now conduct the experiment. Use the table on the next page to help you organize your data.
- Use a two-color chip to simulate the births of 25 children.
- Let one color represent boys and the other color represent girls.
- Toss the chip 25 times.
- After each toss, record the outcome, and compute the ratio of females at that point in the experiment.
- Compute the experimental probability of the outcome at each point in the experiment.

Use graph paper to graph the ordered pairs. (Let x represent the trial number and y represent the experimental probability). Then answer questions 8 and 9.

8. How many female babies were born in 25 births? _____

9. What patterns do you see in the graph? _____
   _____
   _____
   _____

10. Answer question 1 again. Is your answer the same or different? _____
    Explain. _____
    _____
    _____
    _____
    _____

Navigating through Probability in Grades 6–8

Two Hospitals (continued)

Name _____

Trial Number (x-coordinate)	Outcome (M or F)	Ratio of Female Births	Experimental Probability of Female Births (y-coordinate)
1		/1	
2		/2	
3		/3	
4		/4	
5		/5	
6		/6	
7		/7	
8		/8	
9		/9	
10		/10	
11		/11	
12		/12	
13		/13	
14		/14	
15		/15	
16		/16	
17		/17	
18		/18	
19		/19	
20		/20	
21		/21	
22		/22	
23		/23	
24		/24	
25		/25	

Gumball Machine

Name _____

1. Imagine that a gumball machine contains 1 red, 2 green, 3 yellow, and 4 blue gumballs that were thoroughly mixed before they were put into the machine. If you got one gumball from the machine, which color do you think would come out? _____ Explain your answer. _____
   _____
   _____
   _____

2. Suppose that after each gumball comes out, the gumball wizard magically puts another one of the same color into the machine so that the gumball machine always has the same number of each color of gumball. If you took 10 gumballs out of this magic machine, with the gumball wizard replacing your gumball each time, how many times do you think each color would come out? _____
   Explain your answer. _____
   _____
   _____

3. Do the following experiment to simulate the gumball machine:
   - Put 4 blue chips, 3 yellow chips, 2 green chips, and 1 red chip in a paper cup.
   - Mix up the chips thoroughly.
   - Draw one chip out (without looking), tally its color in the middle row of the table below, replace the chip in the cup, and mix the chips again.
   - Repeat these steps until you have tallied 10 trials.

Color of Chip	Blue	Yellow	Green	Red
Tally				
Number				

4. How close were your results to the predictions you made in question 2? _____
   Explain. _____
   _____

5. Do the experiment 10 more times, and record your data below:

Color of Chip	Blue	Yellow	Green	Red
Tally				
Number				

6. Combine the two sets of data in the table below:

Color of Chip	Blue	Yellow	Green	Red
Tally				
Number				

Gumball Machine (continued)

Name _____

7. Using the table in task 6, compute the experimental probability (or relative frequency) of drawing blue, yellow, green, and red:

Color of Chip	Blue	Yellow	Green	Red
Experimental probability				

8. Use B, Y, G, and R to represent the four colors. On the scale below, show the experimental probability of drawing each color:

```
|----|----|----|----|----|----|----|----|----|----|----|----|
0                           1/2                              1
Impossible                                               Certain
```

How Will It Land?

Name _____

1. Tape a penny in the center of the inside bottom of a paper cup. If you toss this cup in the air, what are the possible ways it could land? _____
   _____
   _____

2. Which of the possibilities do you think is most likely? _____
   Explain. _____
   _____

3. Conduct ten trials with your cup. Record each outcome in the second column of the chart below.

Trial	Outcome	Trial	Outcome
1		11	
2		12	
3		13	
4		14	
5		15	
6		16	
7		17	
8		18	
9		19	
10		20	

4. Compute the experimental probability of each outcome. _____
   _____

5. Conduct ten more trials with your cup. Record each outcome in the fourth column of the chart above.

6. Using all twenty trials, compute the experimental probability of each outcome. _____
   _____

7. Are the two sets of experimental probabilities different? _____
   Explain the differences. _____
   _____
   _____

How Will It Land? (continued)

Name _____

8. Tape three more pennies in the center of the inside bottom of the cup. Predict what the new probabilities of the outcomes will be for this changed cup. _____
_____

9. Conduct twenty trials with the new cup, and record the outcomes in the chart below.

Trial	Outcome	Trial	Outcome
1		11	
2		12	
3		13	
4		14	
5		15	
6		16	
7		17	
8		18	
9		19	
10		20	

10. Compute the experimental probability of each outcome. _____
_____
_____

11. Are the probabilities you computed for the four-penny cup different from the probabilities you computed for the cup with only one penny taped inside? _____ Explain the differences.
_____

   How do you account for the differences? _____
_____
_____

The Long Flight Home

Names _____
_____

Four friends—Bob, Bill, Ruth, and Raquel—are among the last to board a jet for a long flight home from camp. Before they board, the flight attendant announces that because of seat limitations, only two of them may sit together; the others must sit separately. The friends decide to hold a drawing to see who will get to sit together. Is it more likely that the two friends who sit together will be the same gender (e.g., "both female") or different genders (e.g., "one male, one female")?

Simulate the selection of the two friends by playing a game. Place two blue cubes and two red cubes in a paper cup, and mix them up. Here are the rules:

- Decide which player is "Same" and which player is "Different."
- Decide which player will start, and then alternate turns.
- On your turn, draw two cubes from the cup without looking.
- If the two cubes are the same color, "Same" scores 1 point. If the two cubes are different colors, "Different" scores 1 point.
- Determine which outcome—"Same" or "Different"—occurred most often after twenty-one draws.

1. Record each draw in the table below:

Turn	Outcome	Turn	Outcome	Turn	Outcome
1		8		15	
2		9		16	
3		10		17	
4		11		18	
5		12		19	
6		13		20	
7		14		21	

2. Which outcome—"Same" or "Different"—occurred more often? _____

3. Simulate the selection of a pair of friends a second time, and record each draw in the table below.

Turn	Outcome	Turn	Outcome	Turn	Outcome
1		8		15	
2		9		16	
3		10		17	
4		11		18	
5		12		19	
6		13		20	
7		14		21	

4. Which outcome—"Same" or "Different"—occurred more often? _____

5. Combine the two sets of results from questions 1 and 3. For the combined data, determine the experimental probabilities of "Same" and "Different." _____

6. Do you think the event "one male, one female" is just as likely as "same gender"? _____
   Explain. _____

Navigating through Probability in Grades 6–8

The Long Flight Home–Extension

Name _____

Imagine that one of the blue cubes has a "1" on it and the other blue cube has a "2" on it; think of these cubes as B_1 and B_2, so B_1 and B_2 are different cubes. Imagine that one of the red cubes has a "1" on it and the other red cube has a "2" on it; think of these cubes as R_1 and R_2, so R_1 and R_2 are also different cubes.

1. List the possible combinations of two cubes that could be drawn from the cup. Use the designations B_1, B_2, R_1, and R_2 for the four cubes.

2. Determine the theoretical probabilities of "same" and "different." _____
   _____
   Do more of the combinations correspond to "same" or "different"? _____

3. Combine your data from question 5 with the data generated by the other students in the class. For the pooled class data, determine the experimental probability of "same" and "different." How do these probabilities compare with the theoretical probabilities you found in question 8? _____
   _____
   _____

4. Which event—"same" or "different"—is more likely? _____ Explain. _____
   _____
   _____

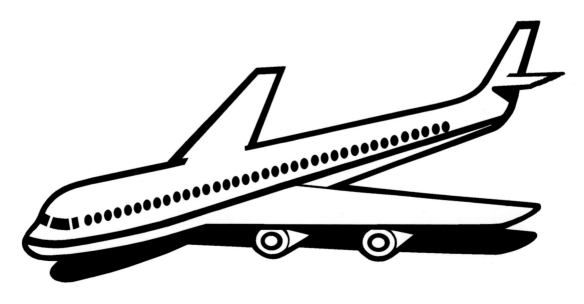

124   Navigating through Probability in Grades 6–8

Dixie's Basketball Contest

Names _____
            _____

Dixie plays on a basketball team. During the season, she made 2/3 of her free throws. At the championship tournament, she has entered the free-throw contest. Each contestant attempts 10 free throws. The person who makes the most baskets is the winner.

Design a simulation to model Dixie's performance. Use the simulation to determine the approximate probability that she will make at least 8 free throws.

1. What is the probability that Dixie will make a basket each time she attempts a free throw? _____
   What is the probability that she will miss a basket each time? _____

2. What device could you use that would generate the probabilities in question 1? Could you use a coin? _____ A die? _____ A spinner? _____
   Another device? _____ Explain your choices. _____
   _____
   _____

3. Do you think that whether Dixie makes a basket on one free throw will affect whether she makes a basket on the next free throw? _____ Explain your answer. _____
   _____

4. Let's assume that Dixie's throws are independent. That is, whether she makes a basket on one throw does not affect whether she makes a basket on any other throw. Explain how you could use one of the devices you identified in question 2 to simulate Dixie's 10 free throws. This process is called a *trial* of the simulation. _____
   _____
   _____

5. Conduct 5 trials, and record the results in the table below.

Trial Number	Results of the Trial	Number of Baskets Made
1		
2		
3		
4		
5		

6. On the basis of your data, determine the approximate probability that Dixie will make at least 8 baskets in 10 free throws. _____

Dixie's Basketball Contest (continued)

Names _____
_____

7. Working with your teacher, pool your data with those of your class. What is the class estimate of the probability that Dixie made at least 8 baskets in 10 free throws? _____

8. Which estimate of the probability do you think is better—the one based on your 5 trials or the one based on the class data? _____ Explain your answer. _____
_____
_____

Newspaper Route

Names _____
_____

Imagine that you have a newspaper route for which you collect $5 per week per customer. One of your customers offers you the following deal: He will give you the $5, or you may draw two bills from a bag that contains one $10 bill and five $1 bills. What should you do? Is this offer a good deal?

1. What are the possible outcomes if you draw two bills from the bag? _____
   Which outcome do you think is more likely, or are they equally likely? _____

2. List two different ways that you could simulate the customer's offer. _____
   Explain how each way accurately models the offer. _____
   _____
   _____
   _____
   _____

3. Choose one of the methods that you listed for question 2. To simulate collecting money according to the customer's offer for 15 weeks, conduct 15 trials of your experiment. Record all the outcomes below.

4. For the 15-week simulation, how many times did you get $2? _____ How many times did you get $11? _____

5. What was the total amount you received for the 15 weeks? _____ What was the mean amount you received each week? _____

6. How much money would you have received in total if you had not drawn money from the bag? _____
   How much would you have received each week? _____

7. Is the customer's offer a good deal? _____ Explain. _____
   _____
   _____

Navigating through Probability in Grades 6–8

How Black Is a Zebra?

Names _____
         _____

Use a random-number table or the random-integer function on a calculator to generate pairs of numbers from 1 through 30. The first number designates the *x*-coordinate of a point, and the second number designates the *y*-coordinate of that point. Determine where each point is on the picture.

- If the point is not on the zebra at all, disregard it, and generate a new pair of random numbers.
- If the point is on the zebra, record the coordinates in the table below. Record ten pairs of coordinates.

Determine if the point is on a black part or a white part of the zebra.

- If it is on a black part, put an *X* in the last column of the table.
- If it is on a white part of the zebra, leave the last column blank.

	First Number (*x*-coordinate)	Second Number (*y*-coordinate)	Black?
1			
2			
3			
4			
5			
6			
7			
8			
9			
10			
		Total Number of Black "Hits"	

1. How many of the recorded points fall on a black part of the zebra? _____ What percent of the points are on a black part? _____

2. Estimate the percent of the zebra that is black. _____

3. How accurate do you think your estimate is? Explain your answer. _____
   _____
   _____

Generate more data points until 20 more points land on the zebra. Tally the number of the new points that "hit" black.

4. How many of the 20 points fell on black? _____ What percent of the points are on black? _____

Navigating through Probability in Grades 6–8

How Black Is a Zebra? (continued)

Names _____
_____

5. Does your set of 20 data points help you make a more accurate estimate of the percent of black on the zebra than your set of 10 data points did? _____ Explain your answer. _____
_____
_____

6. Would you be more confident making an estimate if you had recorded 50 points? _____
100 points? _____ Explain your answer. _____
_____
_____

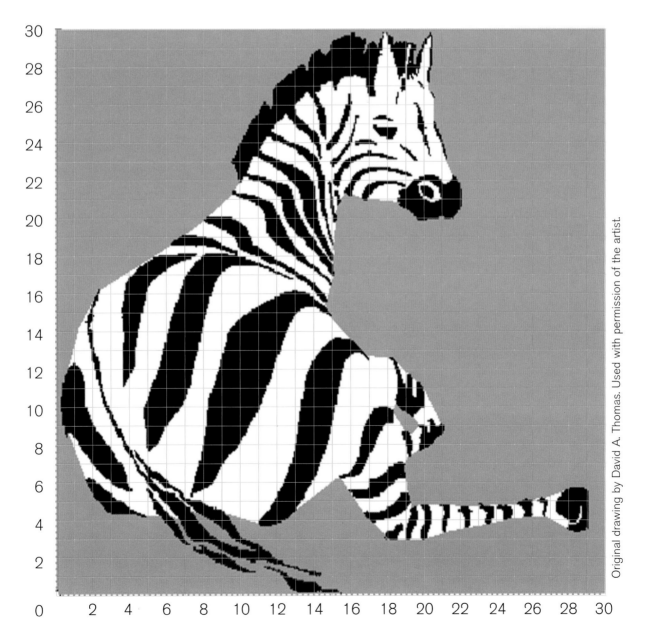

Original drawing by David A. Thomas. Used with permission of the artist.

Solutions for the Blackline Masters

Solutions for "Who Will Win?"

1. The possible outcomes are blue, red, or yellow, but not all outcomes are equally likely.
2. The game is not fair. The spinner is twice as likely to land on blue as on either red or yellow.
3. The possible outcomes are blue, red, or yellow, but not all outcomes are equally likely.
4. The game is not fair. The outcome yellow is less likely than either blue or red.
5. Answers will vary.
6. Answers will vary.

Solutions for "Two Dice"

1. The sums that could result are 2, 3, 4, 5, 6, 7, 8, 9, 10, 11, and 12; the sums are not equally likely; 7 is the most likely, and 2 and 12 are the least likely.
2. The chart displays all the possible pairs and the sums that the pairs would generate.

Red Die	Green Die					
	1	2	3	4	5	6
1	(1, 1) → 2	(1, 2) → 3	(1, 3) → 4	(1, 4) → 5	(1, 5) → 6	(1, 6) → 7
2	(2, 1) → 3	(2, 2) → 4	(2, 3) → 5	(2, 4) → 6	(2, 5) → 7	(2, 6) → 8
3	(3, 1) → 4	(3, 2) → 5	(3, 3) → 6	(3, 4) → 7	(3, 5) → 8	(3, 6) → 9
4	(4, 1) → 5	(4, 2) → 6	(4, 3) → 7	(4, 4) → 8	(4, 5) → 9	(4, 6) → 10
5	(5, 1) → 6	(5, 2) → 7	(5, 3) → 8	(5, 4) → 9	(5, 5) → 10	(5, 6) → 11
6	(6, 1) → 7	(6, 2) → 8	(6, 3) → 9	(6, 4) → 10	(6, 5) → 11	(6, 6) → 12

3. The sums that could result are 2, 3, 4, 5, 6, 7, 8, 9, 10, 11, and 12; the sums are not equally likely; 7 is the most likely, with six pairs that give a sum of 7; 2 and 12 are the least likely; each would occur only once.
4. See the chart in answer 2.
5. The pairs are the same; there is no difference.
6. See the chart. The sums are not equally likely. Sums of 7, 8, 9, 10, 11, 12, 13, and 14 are most likely; those sums are equally likely (the probability of each is 3/36, or 1/12). Sums of 3, 4, 17, and 18 are least likely (the probability of each is 1/12).

First Toss	Second Toss					
	1	2	3	4	5	6
1	(1, 1) → 3	(1, 2) → 4	(1, 3) → 5	(1, 4) → 6	(1, 5) → 7	(1, 6) → 8
2	(2, 1) → 5	(2, 2) → 6	(2, 3) → 7	(2, 4) → 8	(2, 5) → 9	(2, 6) → 10
3	(3, 1) → 7	(3, 2) → 8	(3, 3) → 9	(3, 4) → 10	(3, 5) → 11	(3, 6) → 12
4	(4, 1) → 9	(4, 2) → 10	(4, 3) → 11	(4, 4) → 12	(4, 5) → 13	(4, 6) → 14
5	(5, 1) → 11	(5, 2) → 12	(5, 3) → 13	(5, 4) → 14	(5, 5) → 15	(5, 6) → 16
6	(6, 1) → 13	(6, 2) → 14	(6, 3) → 15	(6, 4) → 16	(6, 5) → 17	(6, 6) → 18

Solutions for "More Often and Most Often"

1. The statement is not true. The probability of drawing a mint is 3/6, or 1/2, and the probability of drawing "not a mint" is also 3/6, or 1/2.
2. The statement is true. The probability of drawing a mint is 3/6, or 1/2; however, the probability of drawing a butterscotch drop is only 2/6, or 1/3, and the probability of drawing a caramel is only 1/6.
3. Mints would be drawn about half the time, so it is not likely that mints would be drawn "most of the time." Students' notions of the meaning of the phrase *most of the time* will influence their answers to this question.
4. The statement is true. Mint (probability of 3/6) or caramel (probability of 1/6) would occur about 3/6 + 1/6 = 4/6, or 2/3, of the time.

Solutions for "Ratios"

1. Answers will vary. See the sample answers in the chart.

Roll	Numbers Rolled and Sum	Prime Sum?	Not a Prime Sum?	Ratio of "Number of Prime Sums" to "Cumulative Number of Rolls"
1	2 + 6 = 8		X	.00
2	4 + 3 = 7	X		.50
3	3 + 4 = 7	X		.67
4	6 + 3 = 9		X	.50
5	3 + 3 = 6		X	.40
6	5 + 6 = 11	X		.50
7	5 + 5 = 10		X	.43
8	3 + 2 = 5	X		.50
9	3 + 6 = 9		X	.44
10	1 + 3 = 4		X	.40
11	3 + 6 = 9		X	.36
12	2 + 2 = 4		X	.33
13	5 + 4 = 9		X	.31
14	4 + 1 = 5	X		.36
15	5 + 2 = 7	X		.40
16	3 + 1 = 4		X	.38
17	4 + 1 = 5	X		.41
18	3 + 1 = 4		X	.39
19	6 + 2 = 8		X	.37
20	6 + 1 = 7	X		.40

2. Graphs will vary. See the graph of the sample data for task 1.

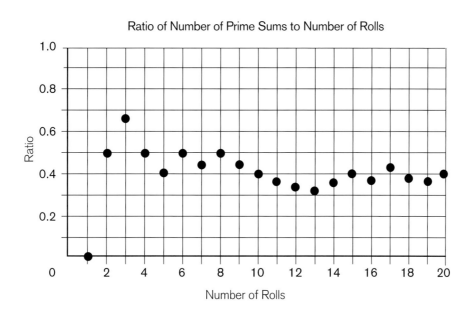

Ratio of Number of Prime Sums to Number of Rolls

3. Answers will vary. As the number of rolls increases, the graph should theoretically approach 42 percent and remain close to that ratio. The sample data are behaving very much as expected.

4. About 42 percent of the sums (2, 3, 5, 7, and 11) would be prime, since five sums are prime and they represent 15/36 (or approximately 42%) of the possible sums.

Navigating through Probability in Grades 6–8

5. Answers will vary. See the sample answers in the chart.

Roll	Numbers Rolled and Product	Two-Digit Product?	Not a Two-Digit Product?	Ratio of "Number of Two-Digit Products" to "Cumulative Number of Rolls"
1	5 × 1 = 5		X	.00
2	2 × 4 = 8		X	.00
3	5 × 2 = 10	X		.33
4	2 × 6 = 12	X		.50
5	4 × 6 = 24	X		.60
6	5 × 4 = 20	X		.67
7	3 × 5 = 15	X		.71
8	2 × 1 = 2		X	.63
9	1 × 1 = 1		X	.56
10	5 × 2 = 10	X		.60
11	1 × 3 = 3		X	.55
12	2 × 2 = 4		X	.50
13	4 × 3 = 12	X		.54
14	6 × 2 = 12	X		.57
15	3 × 6 = 18	X		.60
16	6 × 6 = 36	X		.63
17	1 × 1 = 1		X	.59
18	1 × 3 = 3		X	.56
19	6 × 6 = 36	X		.58
20	2 × 5 = 10	X		.60

6. Graphs will vary. See the graph of the sample data for task 5.

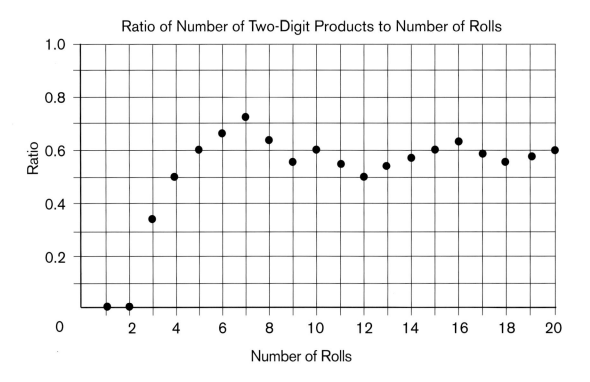

7. Answers will vary. As the number of tosses increases, the graph should theoretically approach 53 percent and remain close to that ratio. The sample data are higher than expected.

8. About 53 percent of the products would be two-digit numbers. Nineteen of the thirty-six possible products are two-digit numbers, so they represent 19/36 (or approximately 53%) of the possible products.

Solutions for "Fair Spinners"

1. See figure 2.6.
2. The range for Pat's spins (18 − 12 = 6) is greater than the range for Chris's spins (11 − 8 = 3), but the number of spins that Pat made was also greater, so it is not clear whether either spinner is fair. (This graph tends to invite an inappropriate comparison of Pat's and Chris's spins instead of focusing attention on the data for either spinner.)
3. See figure 2.7.
4. In the relative-frequency graph, the differences among Pat's bars and Chris's bars are almost the same, so the spinners seem equally fair.
5. The relative-frequency graph shows more clearly that there is little difference between the spinners.

Solutions for "Dice Differences"

1. Answers will vary.
2. Answers will vary.
3. The game is not fair. The chance of A's winning a point is twice the chance of B's winning a point.

Solutions for "Test Guessing"

1. The probability is 1/2 for each question.
2. The eight possible outcomes are CCC, CCW, CWC, CWW, WCC, WCW, WWC, and WWW.
3. The probability of each outcome is 1/8. The guesses are independent, so 1/2 • 1/2 • 1/2 = 1/8. The eight outcomes are equally likely, so each has a probability of 1/8.
4. Only one outcome (CCC) has 70 percent or more correct answers, so the probability is 1/8.
5. For each question, the probability of guessing correctly is 1/2. There are $2^5 = 32$ equally likely outcomes. The thirty-two possible outcomes are CCCCC, CCCCW, CCCWC, CCCWW, CCWCC, CCWCW, CCWWC, CCWWW, CWCCC, CWCCW, CWCWC, CWCWW, CWWCC, CWWCW, CWWWC, CWWWW, WWWWW, WWWWC, WWWCW, WWWCC, WWCCC, WWCCW, WWCWC, WWCWW, WCCCC, WCCCW, WCCWC, WCCWW, WCWCW, WCWCC, WCWWC, and WCWWW. Six of the thirty-two outcomes (CCCCC, CCCCW, CCCWC, CCWCC, CWCCC, WCCCC) have 70 percent or more correct answers, so the probability of passing the test is 6/32.
6. On each question, there are three options, so $P(C) = 1/3$ and $P(W) = 2/3$ for each question. Eight outcomes (CCC, CCW, CWC, WCC, CWW, WCW, WWC, and WWW) are possible, but they are not equally likely:
$$P(CCC) = (1/3)^3 = 1/27$$
$$P(CCW) = P(CWC) = P(WCC) = (1/3) \cdot (1/3) \cdot (2/3) = 2/27$$
$$P(CWW) = P(WCW) = P(WWC) = (1/3) \cdot (2/3) \cdot (2/3) = 4/27$$
$$P(WWW) = (2/3)^3 = 8/27$$
Only one outcome (CCC) has 70 percent or more correct answers: $P(CCC) = (1/3)^3 = 1/27$.

Solutions for "Number Golf: One Die"

Students' scores on the game will vary. The following is an example of how a player might record the play:

Hole 1 (Goal = 15)

Value on Die Tee-Off Number	Cumulative Number
	0
1	1
6	7
2	9
6	15

This player's score is 4, since she reached the goal in four rolls of the die.

Solutions for "Number Golf: Two Dice"

Students' scores on the game will vary. The following is an example of how a player might record the play:

Hole 1 (Goal = 18)

Sum of Values on Dice	Cumulative Number
Tee-Off Number	0
8	8
7	15
11	26
7	33
8	25
5	20
8	12
7	19
5	24
12	12

This player's score is 10, since he was not able to reach the goal in fewer than ten rolls of the dice.

Solutions for "First Head"

For both explorations, the answers will vary. The students may be surprised at the number of tails that occur before the first head.

Solutions for "Strings of Heads"

For all the explorations, the answers will vary.

Solutions for "Racing Game"

1. No; some cars (e.g., 2 and 12) have very small probabilities of winning, whereas others (e.g., 7) have much greater probabilities of winning.
2. Car 7 has the best chance of winning, although cars 6 and 8 have almost as great a chance; the sums 6, 7, and 8 are the most likely to result from the roll of two dice. Cars 2 and 12 are likely to be far behind at the end of the race because the sums 2 and 12 are the least likely to result from the roll of two dice.
3. Answers will vary.
4. Answers will vary.
5. Cars 6, 7, and 8 have the greatest probability of winning because the sums 6, 7, and 8 are most likely to result from the roll of two dice.

Solutions for "Two Hospitals"

1. Choice B is correct. Although both A and B involve at least 80 percent female births, B is more likely because small samples are more likely to produce unexpected results.
2. The unexpected outcome of 80 percent female births is more likely in a small sample because uncharacteristic results are more likely for small samples.
3. Two or three female babies out of five births would be expected; those expected results are close to 50 percent (2/5 = 40% and 3/5 = 60%).
4. Zero, one, four, or five female births in five would be unexpected; those results are not close to 50 percent (0/5 = 0%, 1/5 = 20%, 4/5 = 80%, 5/5 = 100%).
5. Answers will vary.
6. Twelve or thirteen female births would be expected, although any number between ten and fifteen would not be unusual; the expected results are close to 50 percent (12/25 = 48%, 13/25 = 52%).
7. Very small numbers (0–5) or very large numbers (20–25) would be unusual because such extreme values are unlikely.

8. Answers will vary. The table shows sample data. See figure 3.3 for a graph of the sample data.

Trial Number (x-coordinate)	Outcome (M or F)	Ratio of Female Births	Experimental Probability of Female Births (y-coordinate)
1	M	0/1	0
2	F	1/2	.50
3	F	2/3	.67
4	M	2/4	.50
5	M	2/5	.40
6	F	3/6	.50
7	M	3/7	.43
8	M	3/8	.38
9	F	4/9	.44
10	M	4/10	.40
11	F	5/11	.45
12	F	6/12	.50
13	F	7/13	.54
14	M	7/14	.50
15	M	7/15	.47
16	M	7/16	.44
17	F	8/17	.47
18	F	9/18	.50
19	F	10/19	.53
20	F	11/20	.55
21	M	11/21	.52
22	M	11/22	.50
23	F	12/23	.52
24	F	13/24	.54
25	M	13/25	.52

9. As the number of trials increases, the variability in the data would be expected to decrease, resulting in a "flattening out" of the graph.
10. The data from the students' experiments should challenge the reasoning of any student who chose A or who thought the events were equally likely in question 1. The data should indicate that unusual events are more likely in a sample of five events than in a sample of twenty-five events.

Solutions for "Gumball Machine"

1. Blue is more likely than any other color because there are more blue gumballs than gumballs of any other color.
2. Blue would come out about four times; yellow, about three times; green, about two times; and red, about one time. The expected outcomes remain the same because the percent of each color remains constant.

3–7. Answers will vary.

8. The experimental probabilities will vary according to the results of the students' experiments. The theoretical probabilities of the four colors' being drawn are shown on the scale.

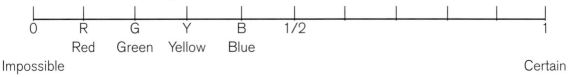

Solutions for "How Will It Land?"

1. The cup could land in one of three ways: upright, upside down, or on its side.
2. Answers will vary. The most likely position is on its side because of the greater area of the sides.
3–6. Answers will vary.
7. The larger data set would be expected to have less variation. Pooling the data from the entire class should result in more-accurate estimates.
8–11. The event "cup lands upright" would be expected to be associated with a higher probability for the four-penny cup than for the one-penny cup. The other outcomes would then have smaller probabilities.

Solutions for "The Long Flight Home"

1. Answers will vary.
2. The outcome "different" is likely to occur more often.
3. Answers will vary.
4. Answers will vary; however, the outcome "different" is likely to occur more often.
5. Answers will vary.
6. The students' precise answers will vary, but most of the experiments will reveal that the outcome "one male, one female" is twice as likely to occur as "same gender."

Solutions for "The Long Flight Home—Extension"

1. The six equally likely outcomes are B_1, B_2; B_1, R_1; B_1, R_2; B_2, R_1; B_2, R_2; R_1, R_2.
2. The probability of "different" is .67, and the probability of "same" is .33. More combinations correspond to "different": four are "different," and two are "same."
3. Answers will vary, but the pooled data would be expected to yield experimental probabilities close to the theoretical values $P(\text{different}) = 2/3$, or .67, and $P(\text{same}) = 1/3$, or .33.
4. The outcome "different" is more likely.

Solutions for "Dixie's Basketball Contest"

1. $P(\text{basket}) = 2/3$; $P(\text{miss}) = 1/3$.
2. A coin is not appropriate, since the probabilities of H and T are each 1/2. A die could be used if two-thirds of the values were designated a "basket" and one-third were designated a "miss"—for instance, basket = {1, 2, 3, 4} and miss = {5, 6}. A spinner would also be appropriate; it could be divided into three equal sectors, with one designating a "miss" and two designating a "basket."
3. The answers will vary. Most people would agree that there might be some slight influence.
4. One example is to roll a die ten times and count the number of times the values designated a "basket"—for instance, 1, 2, 3, or 4—come up; another is to spin a spinner ten times and count the number of times the needle lands on the two sectors designated a "basket."
5. Answers will vary. See the sample data in figure 4.1.
6–7. Answers will vary.
8. The pooled data would be expected to give a better estimate.

Solutions for "Newspaper Route"

1. The possible outcomes are $2 and $11; $2 is more likely.
2. The following are two possible simulations of the situation. Students might also propose a spinner or other appropriate model.

    a) Play money (five $1 bills and one $10 bill) could be put in a bag, and two bills could be drawn out without replacement to model the weekly situation directly. All the bills must be replaced in the bag for subsequent trials.

    b) Alternatively, a die could be rolled, with the numbers 1, 2, 3, 4, and 5 representing $1 and 6 representing $10. On the first draw each week in the actual situation, the carrier would draw one of six bills, and on the second draw, the carrier would draw one of five bills, since one bill would already have been drawn and not replaced.

On the second roll in each simulated trial then, a duplicate of the first number rolled would have to be ignored because in the actual situation, the same bill cannot be drawn twice. So in the simulation with a die,

$$P(\$1 \text{ on first roll}) = \frac{5}{6}$$

and

$$P(\$1 \text{ on second roll}) = \frac{4}{5}.$$

Therefore,

$$P(\$2 \text{ in two rolls}) = P(\$1 \text{ on first roll}) \cdot P(\$1 \text{ on second roll})$$
$$= \frac{5}{6} \cdot \frac{4}{5}$$
$$= \frac{4}{6}$$
$$= \frac{2}{3}.$$

Since the events "$2 in two rolls" and "$11 in two rolls" are complementary (i.e., no other possibilities exist in this situation),

$$P(\$11 \text{ in two rolls}) = 1 - \frac{2}{3}$$
$$= \frac{1}{3}.$$

So the probability of drawing $2 each week is 2/3, and that of drawing $11 each week is 1/3.

3. Answers will vary.
4. Answers will depend on the data the students gathered for question 3.
5. Answers will vary.
6. The total received would have been $75; the weekly payment would have been $5.
7. Students' answers may vary, depending on their data, but if your students are interested in a mathematical analysis of the long-term average value of drawing two bills from the bag, refer to the calculation of the expected value presented on p. 82. One method of calculating 2/3 as the theoretical probability of drawing $2 from the bag each week and 1/3 as the theoretical probability of drawing $11 from the bag each week is given in solution b to question 2. .

Solutions for "How Black Is a Zebra?"

In one simulation, fifteen of thirty-five hits were white, and twenty were black.

1–4. Answers will vary.
5. The twenty-point data set would be expected to be a better sample on which to base an estimate because it is larger.
6. The larger the sample, the more confident the students should be; the larger samples would be expected to yield more-accurate information.

References

Aspinwall, Leslie, and Kenneth L. Shaw. "Enriching Students' Mathematical Intuitions with Probability Games and Tree Diagrams." *Mathematics Teaching in the Middle School* 6 (December 2000): 214–20.

Bird, Elliott. "Counting Attribute Blocks: Constructing Meaning for the Multiplication Principle." *Mathematics Teaching in the Middle School* 5 (May 2000): 568–73.

Brahier, Daniel J. "Genetics as a Context for the Study of Probability." *Mathematics Teaching in the Middle School* 5 (December 1999): 214–21.

Bright, George W., Wallece Brewer, Kay McClain, and Edward S. Mooney. *Navigating through Data Analysis in Grades 6–8*. Reston, Va.: National Council of Teachers of Mathematics, 2003.

Bright, George W., John G. Harvey, and Margariete Montague Wheeler. "Fair Games, Unfair Games." In *Teaching Statistics and Probability*, 1981 Yearbook of the National Council of Teachers of Mathematics, edited by Alfred P. Shulte and James R. Smart, pp. 49–59. Reston, Va.: National Council of Teachers of Mathematics, 1981.

———. *Learning and Mathematics Games*. Journal for Research in Mathematics Education Monograph No. 1. Reston, Va.: National Council of Teachers of Mathematics, 1985.

Bright, George W., and Karl Hoeffner. "Measurement, Probability, Statistics, and Graphing." In *Research Ideas for the Classroom, Middle Grades Mathematics*, edited by Douglas T. Owens, pp. 78–98. New York: Macmillan Publishing Co.; Reston, Va.: National Council of Teachers of Mathematics, 1993.

Dessart, Donald J., and Charlene M. DeRidder. "Readers Write: Is Rock, Scissors, and Paper a Fair Game?" *Mathematics Teaching in the Middle School* 5 (September 1999): 4–5.

Ewbank, William A., and John L. Ginther. "Probability on a Budget." *Mathematics Teaching in the Middle School* 7 (January 2002): 280–83.

Falk, Ruma. "Children's Choice Behaviour in Probabilistic Situations." In *Proceedings of the First International Conference on Teaching Statistics*, edited by D. R. Grey, Peter Holmes, V. Barnett, and G. M. Constable, pp. 714–16. Sheffield, England: Teaching Statistics Trust, 1983.

Fischbein, Efraim. *The Intuitive Sources of Probabilistic Thinking in Children*. Boston: D. Reidel Publishing, 1975.

Freda, Andrew. "Roll the Dice: An Introduction to Probability." *Mathematics Teaching in the Middle School* 4 (October 1998): 85–89.

Jones, Graham A., Carol A. Thornton, Cynthia W. Langrall, and James E. Tarr. "Understanding Students' Probabilistic Reasoning." In *Developing Mathematical Reasoning in Grades K–12*, 1999 Yearbook of the National Council of Teachers of Mathematics, edited by Lee V. Stiff, pp. 146–55. Reston, Va.: National Council of Teachers of Mathematics, 1999.

Kader, Gary, and Mike Perry. "Push-Penny: What Is Your Expected Score?" *Mathematics Teaching in the Middle School* 3 (February 1998): 370–77.

Kahneman, Daniel, Paul Slovic, and Amos Tversky. *Judgment under Uncertainty: Heuristics and Biases*. New York: Cambridge University Press, 1982.

Kahneman, Daniel, and Amos Tversky. "Subject Probability: A Judgment of Representativeness." *Cognitive Psychology* 3 (1972): 430–54.

Konold, Cliff. "Informal Conceptions of Probability." *Cognition and Instruction* 6, no. 1 (1989): 59–98.

National Council of Teachers of Mathematics (NCTM). *Principles and Standards for School Mathematics.* Reston, Va.: NCTM, 2000.

Piaget, Jean, and Bärbel Inhelder. *The Origin of the Idea of Chance in Children.* New York: Norton, 1975.

Scion Corp. Scion Image 1.62c for Mac OS. Frederick, Md.: Scion Corp., n.d.

Shaughnessy, J. Michael. "Research in Probability and Statistics: Reflections and Directions." In *Handbook of Research on Mathematics Teaching and Learning,* edited by Douglas A. Grouws, pp. 465–94. New York: Macmillan Publishing Co.; Reston, Va.: National Council of Teachers of Mathematics, 1992.

Thompson, Denisse R., and Richard A. Austin. "Socrates and the Three Little Pigs: Connecting Patterns, Counting Trees, and Probability." *Mathematics Teaching in the Middle School* 5 (November 1999): 156–61.

Van Zoest, Laura R., and Rebecca K. Walker. "Racing to Understand Probability." *Mathematics Teaching in the Middle School* 3 (October 1997): 162–70.

Wiest, Lynda R., and Robert J. Quinn. "Exploring Probability through an Evens-Odds Dice Game." *Mathematics Teaching in the Middle School* 4 (March 1999): 358–62.

Suggested Reading

Bright, George W. "Game Moves As They Relate to Strategy and Knowledge." *Journal of Experimental Education* 48, no. 3 (1980): 204–9.

Lawrence, Ann. "From *The Giver* to *The Twenty-One Balloons:* Explorations with Probability." *Mathematics Teaching in the Middle School* 4 (May 1999): 504–9.